AT
HOME
IN THE
UNIVERSE

AT
HOME
IN THE
UNIVERSE

The Search
for Laws of
Self-Organization
and Complexity

Stuart Kauffman

New York   Oxford
OXFORD UNIVERSITY PRESS
1995

Oxford University Press

Oxford   New York
Athens   Auckland   Bangkok   Bombay
Calcutta   Cape Town   Dar es Salaam   Delhi
Florence   Hong Kong   Istanbul   Karachi
Kuala Lumpur   Madras   Madrid   Melbourne
Mexico City   Nairobi   Paris   Singapore
Taipei   Tokyo   Toronto

and associated companies in
Berlin Ibadan

Published by Oxford University Press, Inc.,
198 Madison Avenue, New York, New York 10016

Oxford is a registered trademark of Oxford University Press

Library of Congress Cataloging-in-Publication Data
Kauffman, Stuart A.
At home in the universe: the search for laws of self-organization
and complexity / Stuart Kauffman.
p. cm.   Includes bibliographical references and index.
ISBN 0-19-509599-5
1. Life—Origin.   2. Self-organizing systems.   3. Molecular
evolution.   4. Evolution—Philosophy.   I. Title.
QH325.K388 1995
577—dc20 94-25268

9 8 7 6 5 4 3 2 1

Printed in the United States of America
on acid-free paper

To my colleagues
at the Santa Fe Institute
and elsewhere,
who, like me,
search for the
laws of complexity.

We live in a world of stunning biological complexity. Molecules of all varieties join in a metabolic dance to make cells. Cells interact with cells to form organisms; organisms interact with organisms to form ecosystems, economies, societies. Where did this grand architecture come from? For more than a century, the only theory that science has offered to explain how this order arose is natural selection. As Darwin taught us, the order of the biological world evolves as natural selection sifts among random mutations for the rare, useful forms. In this view of the history of life, organisms are cobbled-together contraptions wrought by selection, the silent and opportunistic tinkerer. Science has left us as unaccountably improbable accidents against the cold, immense backdrop of space and time.

Thirty years of research have convinced me that this dominant view of biology is incomplete. As I will argue in this book, natural selection is important, but it has not labored alone to craft the fine architectures of the biosphere, from cell to organism to ecosystem. Another source—self-organization—is the root source of order. The order of the biological world, I have come to believe, is not merely tinkered, but arises naturally and spontaneously because of these principles of self-organization—laws of complexity that we are just beginning to uncover and understand.

The past three centuries of science have been predominantly reductionist, attempting to break complex systems into simple parts, and those parts, in turn, into simpler parts. The reductionist program has been spectacularly successful, and will continue to be so. But it has often left a vacuum: How do we use the information gleaned about the parts to build up a theory of the whole? The deep difficulty here lies in the fact that the complex whole may exhibit properties that are not readily explained by understanding the parts. The complex whole, in a completely nonmystical sense, can often exhibit collective properties, "emergent" features that are lawful in their own right.

This book describes my own search for laws of complexity that govern how life arose naturally from a soup of molecules, evolving into the biosphere we see today. Whether we are talking about molecules cooperating to form cells or organisms cooperating to form ecosystems or buyers and sellers cooperating to form markets and economies, we will find grounds to believe that Darwinism is not enough, that natural selection cannot be the sole source of the order we see in the world. In crafting the living world, selection has always acted on systems that exhibit spontaneous order. If I am right, this underlying order, further honed by selection, augurs a new place for us—expected, rather than vastly improbable, at home in the universe in a newly understood way.

Santa Fe, N.M.   S. K.
October 1994

ACKNOWLEDGMENTS

This book could not have been written without the wise editorial crafting of George Johnson, a fine science writer and *New York Times* editor whose own book, *Fire in the Mind: Science, Faith, and the Search for Order*, was written even as he was helping me with *At Home in the Universe*. Early lunches on La Posada's patio in Santa Fe allowed us to lay out the bare bones. I was glad to offer George co-authorship. He refused, feeling that *At Home in the Universe* should be my book, one mind's transect through the emerging sciences of complexity. This was not the last time I took George's counsel. Every week or two, we would meet at La Posada or hike the high-hill backdrop of Santa Fe, two ranges back from the prairies to avoid Apache raids, and talk through each chapter one by one. In return, I introduced George to the joy of mushrooming—he managed to find more lovely boletes than I on our first try. Then it was home to my office to sit in front of the keyboard and, in the standard outpouring of enthusiasm and agony all writers know, spill a chapter onto the electronic page. I would give the result to George, who carved, coaxed, laughed, and hewed. After six such cycles, as I responded each time to his polite demands to cut this, expand that, I was delighted to see a book emerge. The writing is mine, the voice is mine, but the book has a clarity and structure I would not have achieved alone. I am honored to have had George's help.

CONTENTS

1. At Home in the Universe, 3

2. The Origins of Life, 31

3. We the Expected, 47

4. Order for Free, 71

5. The Mystery of Ontogeny, 93

6. Noah's Vessel, 113

7. The Promised Land, 131

8. High-Country Adventures, 149

9. Organisms and Artifacts, 191

10. An Hour upon the Stage, 207

11. In Search of Excellence, 245

12. An Emerging Global Civilization, 273

BIBLIOGRAPHY, 305

INDEX, 307

AT
HOME
IN THE
UNIVERSE

Chapter 1

At Home in
the Universe

Out my window, just west of Santa Fe, lies the near spiritual land-scape of northern New Mexico—barrancas, mesas, holy lands, the Rio Grande—home to the oldest civilization in North America. So much, so ancient and modern, pregnant with the remote past and the next millennium mingle here, haphazardly, slightly drunk with anticipation. Forty miles away lies Los Alamos, brilliance of mind, brilliance of flashing light that desert dawning in 1945, half a century ago, half our assumptions ago. Just beyond spreads the Valle Grande, remains of an archaic mountain said to have been over 30,000 feet high that blew its top, scattering ash to Arkansas, leaving obsidian for later, finer work-ings.

Some months ago, I found myself at lunch with Gunter Mahler, a theoretical physicist from Munich visiting the Santa Fe Institute, where a group of colleagues and I are engaged in a search for laws of complex-ity that would explain the strange patterns that spring up around us. Gunter looked northward, past piñon and juniper, taking in the long view toward Colorado, and somewhat astonished me by asking what my image of paradise was. As I groped for an answer, he proposed one: not the high mountains, or the ocean's edges, or flat lands. Rather, he sug-gested, just such terrain as lay before us, long and rolling under strong light, far ranges defining a distant horizon toward which graceful and telling land forms march in fading procession. For reasons I do not completely understand, I felt he was right. We soon fell to speculations about the landscape of East Africa, and wondered whether, in fact, we might conceivably carry some genetic memory of our birthplace, our real Eden, our first home.

What stories we tell ourselves, of origins and endings, of form and transformation, of gods, the word, and law. All people, at all times, must have created myths and stories to sketch a picture of our place under the sun. Cro-Magnon man, whose paintings of animals seem to exhibit a respect and awe, let alone line and form, that equals or surpasses those of later millennia, must have spun answers to these questions: Who are we? Where did we come from? Why are we here? Did Neanderthal, *Homo habilis,* or *Homo erectus* ask? Around which fire in the past 3 million years of hominid evolution did these questions first arise? Who knows.

Somewhere along our path, paradise has been lost, lost to the Western mind, and in the spreading world civilization, lost to our collective mind. John Milton must have been the last superb poet of Western civilization who could have sought to justify the ways of God to man in those early years foreshadowing the modern era. Paradise has been lost, not to sin, but to science. Once, a scant few centuries ago, we of the West believed ourselves the chosen of God, made in his image, keeping his word in a creation wrought by his love for us. Now, only 400 years later, we find ourselves on a tiny planet, on the edge of a humdrum galaxy among billions like it scattered across vast megaparsecs, around the curvature of space–time back to the Big Bang. We are but accidents, we're told. Purpose and value are ours alone to make. Without Satan and God, the universe now appears the neutral home of matter, dark and light, and is utterly indifferent. We bustle, but are no longer at home in the ancient sense.

We accept, of course, that the rise of science and the consequent technological explosion has driven us to our secular worldview. Yet a spiritual hunger remains. I recently met N. Scott Momaday, a Native American author, winner of the Pulitzer Prize, at a small meeting in northern New Mexico intended to try to articulate the fundamental issues facing humanity. (As if a small group of thinkers could possibly succeed.) Momaday told us that the central issue we confront is to reinvent the sacred. He told of a sacred shield of the Kiowa, sanctified by the sacrifices and suffering of the warriors who had been honored to hold it in battle. The shield had been stolen following a battle with U.S. cavalry forces after the Civil War. He told us of the recent discovery and return of the shield from the family home of the post–Civil War general who had taken it. Momaday deep voice fell gently over us as he described the welcome set forth for that shield and the place it holds, quiet, somber, still, revered for the passion and suffering distilled within its arc.

Momaday's search for the sacred settled deep on me, for I hold the hope that what some are calling the new sciences of complexity may help us find anew our place in the universe, that through this new science, we may recover our sense of worth, our sense of the sacred, just as the Kiowa ultimately recovered that sacred shield. At the same meeting, I suggested that the most important problem confronting humanity was the emergence of a world civilization, its profound promise, and the cultural dislocations this transformation will cause. To undergird the pluralistic global community that is aborning, we shall need, I think, an expanded intellectual basis—a new way to think about origins, evolution, and the profound naturalness of life and its myriad patterns of unfolding. This book is an effort to contribute to that new view, for the emerging sciences of complexity, as we shall see, offer fresh support for the idea of a pluralistic democratic society, providing evidence that it is not merely a human creation but part of the natural order of things. One is always wary of deducing from first principles the political order of one's own society. The nineteenth-century philosopher James Mill once succeeded in deducing from first principles that a constitutional monarchy, remarkably like that in England early in the last century, was the highest natural form of governance. But, as I hope to show, the very laws of complexity my colleagues and I are seeking suggest that democracy has evolved as perhaps the optimal mechanism to achieve the best attainable compromises among conflicting practical, political, and moral interests. Momaday must be right as well. We shall also need to reinvent the sacred—this sense of our own deep worth—and reinvest it at the core of the new civilization.

The story of our loss of paradise is familiar but worth retelling. Until Copernicus, we believed ourselves to be at the center of the universe. Nowadays, in our proclaimed sophistication, we look askance at a church that sought to suppress a heliocentric view. Knowledge for knowledge's sake, we say. Yes, of course. But was the church's concern with the disruption of a moral order really no more than a narrow vanity? To pre-Copernican Christian civilization, the geocentric view was no mere matter of science. Rather, it was the cornerstone evidence that the entire universe revolved around us. With God, angels, man, the beasts, and fertile plants made for our benefit, with the sun and stars wheeling overhead, we knew our place: at the center of God's creation. The church feared rightly that the Copernican views would ultimately dismantle the unity of a thousand-year-old tradition of duty and rights, of obligations and roles, of moral fabric.

Copernicus blew his society open. Galileo and Kepler did not help much either, particularly Kepler, with his demonstration that planets orbit in ellipses rather than in the rational perfect circles envisioned by Aristotle. Kepler is such a wonderful transitional figure, a descendant of the tradition of the magus, or great magi a century earlier. He had not sought ellipses, but rather harmonic orbits corresponding to the five perfect solids, from which Plato himself tried to build the world.

Then Newton, hero to us all, escaped the plague and wrenched Everyman into a universe even farther from paradise. What a giant step he took. Just imagine what it must have felt like to Newton as his new laws of mechanics took form in his mind. What wonder he must have felt. With a mere three laws of motion and a universal law of gravitation in hand, Newton not only derived tides and orbits, but unleashed on the Western mind a clockwork universe. Before Newton, a scholastic philosopher, certain that an arrow arced toward its target because, as Aristotle taught, it was constantly acted on by a mysterious force, or impetus, could easily believe in a God who also moved things along by according them his sustained attention. Such a God might look after one if properly addressed. Such a God might return one to paradise. But after Newton, the laws alone sufficed. The universe could be wound up by God and released, thereafter left to tick inevitably toward eternity under the unfolding of his law, without further intervention. If the stars and tides moved without divine intervention, thinking people began to find it more difficult to hope for such intervention in their own affairs.

But there was some consolation. If the planets and all of inanimate matter obey eternal laws, then surely living things, with man at their summit, sitting at the top of the great chain of being, must reflect God's intention. Adam himself had named them all: insects, fish, reptiles, birds, mammals, man. Like the hierarchy of the church itself—laity, priest, bishop, archbishop, pope, saints, angels—the great chain of being stretched from the lowliest to the Almighty.

How Darwin and his theory of evolution by natural selection devastated all this! We, who are heritors of Darwin, who see the living world through the categories he taught us more than a century ago, even we have trouble with the implications: man as the result of a chain of accidental mutations, sifted by a law no more noble than survival of the fittest. Creation science is no accident in late-twentieth-century America. Its proponents adhere to it in an ardent effort to forestall the feared moral implications of humans as descendants of a haphazard lineage branching from some last common ancestor more remote in time than the Cambrian explosion some 500 million years ago. The science in creation science is no science at all, but is the moral anguish so foolish? Or should creationism be viewed rather more sympathetically—misguided, to be sure, but part of a broader quest to reinvent the sacred in our secular world?

Before Darwin, the so-called Rational Morphologists found comfort in the view that species were not the result of random mutation and selection, but of timeless laws of form. The finest eighteenth- and early-nineteenth-century biologists, comparing living forms, classified them into the hierarchical groupings of Linnean taxonomy that remain with us today: species, genera, families, orders, classes, phyla, and kingdoms—an order as natural and cosmically ordained to the scientists of that time as the great chain of being was to the Catholic church. Given the obvious morphological overlaps, these scientists sought lawful explanations of the similarities and differences. An analogy that helps us understand this aim can be found by thinking of crystals, which can exist only in certain forms. The Rational Morphologists, confronting fixed but highly similar species, sought similar regularities. Surely the pectoral fin of a fish, the bones of a petrel's wings, and the flashing legs of a horse were expressions of the same deep principle.

Darwin devastated this world. Species are not fixed by the squares of the Linnean chart; they evolve from one another. Natural selection acting on random variation, not God or some principle of Rational Morphology, accounted for the similarity of limb and fin, for creatures so marvelously attuned to their environments. The implications of these ideas, as understood by biologists today, have transformed humans, along with all other living forms, from works of a creator into ultimately historical accidents wrought by the opportunism of evolution. "Evolution is chance caught on the wing," wrote the biologist Jacques Monod. Evolution tinkers together contraptions, agreed François Jacob, the French geneticist who shared the Nobel with Monod. In the Judeo-Christian tradition, we had become used to thinking of ourselves as fallen angels. At least fallen angels have some hope of redemption and grace, of climbing back up the ecclesiastical ladder. Evolution left us stuck on the earth with no ladder to climb, contemplating our fate as nature's Rube Goldberg machines.

Random variation, selection sifting. Here is the core, the root. Here lies the brooding sense of accident, of historical contingency, of design by elimination. At least physics, cold in its calculus, implied a deep order, an inevitability. Biology has come to seem a science of the accidental, the ad hoc, and we just one of the fruits of this ad hocery. Were the tape played over, we like to say, the forms of organisms would surely differ dramatically. We humans, a trumped-up, tricked-out, horn-blowing, self-important presence on the globe, need never have occurred. So much for our pretensions; we are lucky to have our hour. So much, too, for paradise.

Where, then, does this order come from, this teeming life I see from my window: urgent spider making her living with her pre-nylon web, coyote crafty across the ridgetop, muddy Rio Grande aswarm with no-see-ems (an invisible insect peculiar to early evenings)? Since Darwin, we turn to a single, singular force, Natural Selection, which we might as well capitalize as though it were the new diety. Random variation, selection-sifting. Without it, we reason, there would be nothing but incoherent disorder.

I shall argue in this book that this idea is wrong. For, as we shall see, the emerging sciences of complexity begin to suggest that the order is not all accidental, that vast veins of spontaneous order lie at hand. Laws of complexity spontaneously generate much of the order of the natural world. It is only then that selection comes into play, further molding and refining. Such veins of spontaneous order have not been entirely unknown, yet they are just beginning to emerge as powerful new clues to the origins and evolution of life. We have all known that simple physical systems exhibit spontaneous order: an oil droplet in water forms a sphere; snowflakes exhibit their evanescent sixfold symmetry. What is new is that the range of spontaneous order is enormously greater than we have supposed. Profound order is being discovered in large, complex, and apparently random systems. I believe that this emergent order underlies not only the origin of life itself, but much of the order seen in organisms today. So, too, do many of my colleagues, who are starting to find overlapping evidence of such emergent order in all different kinds of complex systems.

The existence of spontaneous order is a stunning challenge to our settled ideas in biology since Darwin. Most biologists have believed for over a century that selection is the sole source of order in biology, that selection alone is the "tinkerer" that crafts the forms. But if the forms selection chooses among were generated by laws of complexity, then selection has always had a handmaiden. It is not, after all, the sole source of order, and organisms are not just tinkered-together contraptions, but expressions of deeper natural laws. If all this is true, what a revision of the Darwinian worldview will lie before us! Not we the accidental, but we the expected.

The revision of the Darwinian worldview will not be easy. Biologists have, as yet, no conceptual framework in which to study an evolutionary process that commingles both self-organization and selection. How does selection work on systems that already generate spontaneous order? Physics has its profound spontaneous order, but no need of selection. Biologists, subliminally aware of such spontaneous order, have nevertheless ignored it and focused almost entirely on selection. Without a framework to embrace both self-organization and selection, self-organization has been rendered almost invisible, like the background in a gestalt picture. With a sudden visual shift, the background can become the foreground, and the former foreground, selection, can become the background. Neither alone suffices. Life and its evolution have always depended on the mutual embrace of spontaneous order and selection's crafting of that order. We need to paint a new picture.

Genesis

Two conceptual lineages from the nineteenth century linked to complete our sense of accidental isolation in the swirl of stars. In addition to Darwin, the science of thermodynamics and statistical mechanics, from the French engineer Sadi Carnot to the physicists Ludwig Boltzmann and Josiah Willard Gibbs, gave us the seemingly mysterious second law of thermodynamics: In equilibrium systems—those closed to the exchange of matter and energy with their environment—a measure of disorder called entropy inevitably increases. We have all seen the simple examples. A droplet of dark blue ink in a dish of still water diffuses to a uniform light blue. The ink does not reassemble into a single droplet.

Boltzmann gave us the modern understanding of the second law. Consider a box filled with gas molecules, modeled as hard elastic spheres. All the molecules could be in a small corner of the box, or they could be spread out more or less uniformly. One arrangement is just as unlikely as any other. But a vastly larger number of the possible configurations correspond to situations in which the molecules are more or less uniformly distributed than to situations with all the molecules confined, say, to a single corner. Boltzmann argued that the increase in entropy in equilibrium systems arises from nothing more than the statistical tendency of the system to pass randomly through all possible arrangements (the so-called ergodic hypothesis). In the vast majority of these cases, the molecules will be distributed uniformly. And so, on average, that is what we will see. The ink droplet diffuses and does not reassemble; the gas molecules diffuse from one corner of the box and do not reassemble. Left to itself, a system will visit all possible microscopic, fine-grained configurations equally often. But the system will spend most of its time in those coarse-grained patterns satisfied by very large numbers of fine-grained patterns—molecules uniformly distributed throughout the box. So the second law is not so mysterious, after all.

The consequence of the second law is that in equilibrium systems, order—the most unlikely of the arrangements—tends to disappear. If order is defined as those coarse-grained states that correspond to only a few fine-grained states (molecules bunched in the upper-left-hand corner, molecules arranged in a plane parallel to the top of the box), then at thermodynamic equilibrium, those delicate arrangements disappear because of the ergodic wandering of the system through all its microstates. It follows that the maintenance of order requires that some form of work be done on the system. In the absence of work, order disappears. Hence we come to our current sense that an incoherent collapse of order is the natural state of things. Again, we the accidental, we the unexpected.

The second law of thermodynamics has been thought to be rather gloomy. One almost imagines the grave headlines: UNIVERSE RUNNING DOWN. HEAT DEATH HEADED OUR WAY. DISORDER IS ORDER OF THE DAY. How far we have come from the blessed children of God, at the center of the universe, walking among creatures created for our benefit, in a garden called Eden. Science, not sin, has indeed lost us our paradise.

If the universe is running down because of the second law, the easy evidence out my window is sparse—some litter here and there, and the heat given off by me, a homeotherm, scrambling the molecules of air. It is not entropy but the extraordinary surge toward order that strikes me. Trees grabbing sunlight from a star eight light-minutes away, swirling its photons together with mere water and carbon dioxide to cook up sugars and fancier carbohydrates; legumes sucking nitrogen from bacteria clinging to their roots to create proteins. I eagerly breathe the waste product of this photosynthesis, oxygen—the worst poison of the archaic world, when anaerobic bacteria ruled—and give off the carbon dioxide that feeds the trees. The biosphere around us sustains us, is created by us, grafts the energy flux from the sun into the great web of biochemical, biological, geologic, economic, and political exchanges that envelopes the world. Thermodynamics be damned. Genesis, thank whatever lord may be, has occurred. We all thrive.

The earliest signs of life on earth are present 3.45 billion years ago, 300 million years after the crust cooled sufficiently to support liquid water. These signs of life are hardly trivial. Well-formed cells, or what the experts believe are cells, are present in the archaic rocks from that period. Figure 1.1 shows such ancient fossils. Figure 1.2a shows contemporary coccoid cyanobacteria, while Figure 1.2b shows similar fossil cyanobacteria from 2.15 billion years ago. The morphological similarity is stunning. These ancient cells appear to have a cell membrane that separates an internal milieu from an external environment. Of course, morphological similarity does not demonstrate biochemical or metabolic similarity, but we look at these fossils with the chilling feeling that we are seeing imprints of the universal ancestor.

Cells were undoubtedly the triumphant culmination of some form of

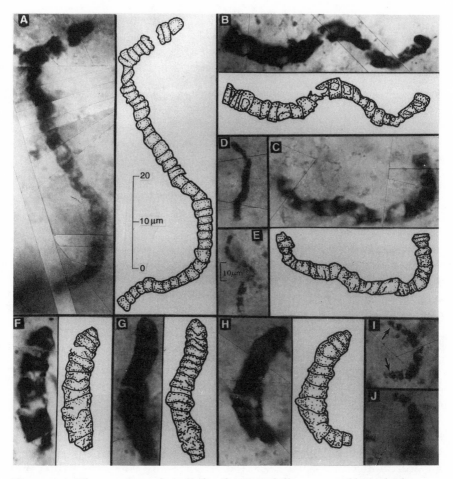

Figure 1.1 *The ancestors of us all: fossils, 3.437 billion years old. Each photograph is accompanied by an artist's rendition.*

evolution from the earliest webs of interacting molecules complex enough to exhibit the qualities we call alive: the ability to metabolize, to reproduce, and to evolve. In turn, the origin of precellular life was itself the triumphant culmination of some form of prebiotic chemical evolution, which led from the limited diversity of molecular species present in the gas cloud of the protoearth to the enhanced chemical diversity that underwrote the crystallization of life, of self-reproducing molecular systems.

But once cells were invented, a numbing lassitude befell our earliest ancestors. Something like simple single-celled organisms, presumably related to current archaebacteria, and later, elongated tubelike cells that presumably were precursors of early fungi, seem to have formed a

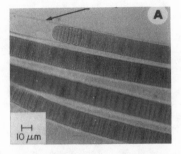

Figure 1.2 *Ancient fossil cells and modern living cells show a stunning resemblance.* (a) *Living coccoid cyanobacteria colony.* (b) *Fossil coccoid cyanobacteria colony, 2.15 billion years old, from Canada.*

global co-prosperity sphere that persisted for perhaps 3 billion rather dull years. These progenitors of all later ecosystems formed complex local economies in which bacterial species and algae, competing and collaborating with one another, formed complex mound structures, typically several meters across and several meters high. Contemporary mounds are found in abundance along the Great Barrier Reef on the northeastern coast of Australia. Fossilized versions of these mounds are called stromatoliths. Presumably, these simple ecosystems covered the shallow coastal waters of the globe. Even today, the same type of formations are found in shallow waters in the Gulf of California and Australia. Contemporary mounds harbor hundreds of bacterial species and a modest number of algal species. One guesses that archaic mound systems were of a similar complexity.

Such single-celled life-forms persisted alone in the biosphere for perhaps 3 billion years—most of the age of the earth. But life-forms were destined to change by some unknown agency, some pregnancy of possibilities. Was it Darwinian chance and selection alone, as biologists have argued for decades? Or, as I shall propose in this book, did principles of self-organization mingle with chance and necessity? By perhaps 800 million years ago, multicellular organisms appeared. The routes to the formation of such multicellular life remain obscure, although some workers believe that it occurred when a tubelike protofungi began to form internal walls that perhaps later became individual cells.

All hell broke loose in the Cambrian explosion, about 550 million years ago. A burst of evolutionary creativity generated almost all the major phyla now jostling one another in nooks and crannies on, above, under the earth's surface, everywhere, even thousands of feet down in solid rock. Only the vertebrates, our own lineage, arose a bit later, in Ordovician times.

The history of life in the first 100 million years following the Cambrian explosion was one of bustling confusion. And its mysteries are yet to be solved. The Linnean chart groups organisms hierarchically, from the specific to the general: species, genera, families, orders, classes, phyla, and kingdoms. One might imagine that the first multicellular creatures would all be very similar, only later diversifying, from the bottom up, into different genera, families, orders, classes, and so on. That, indeed, would be the expectation of the strictest conventional Darwinist. Darwin, profoundly influenced by the emerging view of geologic gradualism, proposed that all evolution occurred by the very gradual accumulation of useful variations. Thus the earliest multicellular creatures themselves ought to have diverged gradually from one another. But this appears to be false. One of the wonderful and puzzling features of the Cambrian explosion is that the chart was filled in from the top down. Nature suddenly sprang forth with many wildly different body plans—the phyla—elaborating on these basic designs to form the classes, orders, families, and genera.

In his book about the Cambrian explosion, *Wonderful Life: The Burgess Shale and the Nature of History,* Stephen Jay Gould remarks on this top–down quality of the Cambrian with wonder. In the Permian extinction, 245 million years ago, 96 percent of all species disappeared. But in the rebound, during which many new species evolved, diversity filled in from the bottom up, with many new families, a few new orders, one new class, and no new phyla.

Many are the arguments about this asymmetry between the Cambrian and the Permian explosions. My own view, explored in later chapters, is that the Cambrian explosion is like the earliest stages of the technological evolution of an entirely novel invention, such as the bicycle. Recall the funny early forms: big front wheels, little back ones; little front wheels, big back ones. A flurry of forms branched outward over Europe and the United States and elsewhere, giving rise to major and minor

variants. Soon after a major innovation, discovery of profoundly different variations is easy. Later innovation is limited to modest improvements on increasingly optimized designs.

So it was, I think, in the Cambrian, when multicellular life first tested
out its possible modes of being. The particular branchings of life, were
the tape played again, might differ, but the *patterns* of the branching,
dramatic at first, then dwindling to twiddling with details later, are
likely to be lawful. Biological evolution may be a deeply historical
process, as Darwin taught us, but lawlike at the same time.

As we shall see in later chapters, the parallels between branching evolution in the tree of life and branching evolution in the tree of technology bespeak a common theme: both the evolution of complex organisms and the evolution of complex artifacts confront conflicting "design
criteria." Heavier bones are stronger, but may make agile flight harder
to achieve. Heavier beams are stronger, but may make agile fighter aircraft harder to achieve as well. Conflicting design criteria, in organism
or artifact, create extremely difficult "optimization" problems—jug-
gling acts in which the aim is to find the best array of compromises. In
such problems, crudely designed major innovations can be greatly improved by dramatic variations on the new theme. Later on, as most of
the major innovations have been tried, improvements dwindle to mere
fiddling with the details. If something like this is true, then evolution's
cadences may find echos in the evolution of artifacts and cultural forms
we human tinkerers create.

The past 550 million years have witnessed well-fossilized life-forms
emerging onto and then ebbing from the stage. Speciation and extinction go roughly hand in hand. Indeed, recent evidence suggests that the
highest rate of extinction, as well as speciation, occurred in the Cambrian itself. Over the next 100 million years, the average diversity of
species increased to a kind of rough steady state. But that level was, and
persistently is, perturbed by small and large avalanches of extinctions
that wipe out modest numbers or large numbers of species, genera, or
families. Many of these catastrophes may have been caused by small and
large meteors. Indeed, the extinction at the end of the Cretaceous,
which coincided with the denouement of the dinosaurs, was probably
caused by a massive misfortune that landed near the Yucatán.

In this book, I shall explore a different possibility. It does not always
take a meteor or some outside cataclysm to wipe out whole species.
Rather, speciation and extinction seem very likely to reflect the spontaneous dynamics of a community of species. The very struggle to survive,
to adapt to the small and large changes of one's coevolutionary partners,
may ultimately drive some species to extinction while creating novel

niches for others. Life, then, unrolls in an unending procession of change, with small and large bursts of speciations, and small and large bursts of extinctions, ringing out the old, ringing in the new. If this view is correct, then the patterns of life's bursts and burials are caused by internal processes, endogenous and natural. These patterns of speciations and extinctions, avalanching across ecosystems and time, are somehow self-organized, somehow collective emergent phenomena, somehow natural expressions of the laws of complexity we seek. And somehow, when understood, such patterns must afford us a deeper understanding of the game we have all joined, for we are all part of the same pageant.

No small matter these small and large avalanches of creativity and destruction, for the natural history of life for the past 550 million years has echoes of the same phenomena at all levels: from ecosystems to economic systems undergoing technological evolution, in which avalanches of new goods and technologies emerge and drive old ones extinct. Similar small and large avalanches even occur in evolving cultural systems. The natural history of life may harbor a new and unifying intellectual underpinning for our economic, cultural, and social life. I will spend much of this book unpacking the grounds to think that a deep theory for such ceaseless change can be found. I suspect that the fate of all complex adapting systems in the biosphere—from single cells to economies—is to evolve to a natural state between order and chaos, a grand compromise between structure and surprise. Here, at this poised state, small and large avalanches of coevolutionary change propagate through the system as a consequence of the small, best choices of the actors themselves, competing and cooperating to survive. I will suggest that, on small and large scales, we all do the best we can but will eventually be hustled offstage by some unanticipated consequences of our own best efforts. We will find a place in the sun, poised on the edge of chaos, sustained for a time in that sun's radiance, but only for a moment before we slip from sight. Untold many actors come and go, each, as a fine playwright once said, strutting and fretting its hour upon the stage. A smiling irony is our fate.

We all make our livings—frog, fern, bracken, bird, seafarer, or landed gentry. From the metabolic mutualisms of legume root and nitrogen-fixing bacteria, by which each makes a nutrient needed by the other, to the latest research partnership between drug giant and small biotech firm, we are all selling and trading our stuff to one another to get our daily bread. And somehow, the burgeoning diversity of the Cambrian— where each new species offers a novel niche or two to the others that feed on it, flee from it, or share with it—looks rather like the burgeoning diversity of an economic system in which each new good or service affords a niche or two for other goods or services, whose providers thereby make a living. We are all trading our stuff to one another. We all must make our living. Might general laws govern all this activity? Might general laws govern phenomena ranging from the Cambrian explosion to our postmodern technological era, in which the exploding rate of innovation brings the time horizon of future shock ever closer? That is the possibility I will be exploring in this book.

The Laws of Life

Whence cometh all this activity, complexity, and exuberant chutzpah? If the physicists are right, all this diversity can only be understood as a consequence of the fundamental laws they have sought since Kepler and Galileo started becoming too advanced for the church. This most profound of scientific hopes, this search for fundamental laws, is to be held in the deepest respect: it is the ideal of reductionism in science. As Steven Weinberg phrased it in his recent book title, it is the "dream of a final theory." Weinberg's characterization of the ancient quest is heartfelt. We seek reductionist explanations. Economic and social phenomena are to be explained in terms of human behavior. In turn, that behavior is to be explained in terms of biological processes, which are in turn to be explained by chemical processes, and they in turn by physical ones.

Much has been said about the validity of the reductionist program. But this much all of us would agree on: if we should find the final theory—perhaps superstrings embedded in 10-dimensional space with 6 dimensions curled in on themselves, and the remaining 4 whipped into some topological foam of quantized space–time, allowing gravity and the other three forces to fit into one conceptual framework—if we should find the final theory, we should have only begun our task. For on that genuinely magnificent day when the fundamental law should be graven forever on some Carrara marble stele or, as the physicist Leon Lederman suggests, on the front of a T-shirt, we should then truly have to begin to compute the consequences of that law.

Could we ever hope to carry out this second half of the reductionist program? Could we use the laws to understand the biosphere we see? We confront here the distinction between explaining and predicting. A table of the tides predicts but does not explain. Newton's theory predicts and explains. Many biologists think that Darwin's theory explains but only weakly predicts. Our final theory of physics might well explain, but almost certainly will not predict in detail. Failure to predict can itself already be predicted on at least two grounds. The first is quantum mechanics, which ensures a fundamental indeterminism at the subatomic level. Since that indeterminism has macroscopic consequences—for example, a random quantum event can cause a mutation in DNA molecules—we appear to be fundamentally forbidden from deriving detailed specific predictions about all molecular and supramolecular events. The second difficulty derives from the mathematical field now known as chaos theory. The central idea is simple and is captured in the so-called butterfly effect: a legendary butterfly flapping its wings in Rio changes the weather in Chicago. (I have lived in Chicago and personally suspect that nothing can change the weather there.) It always appears to be the same butterfly whenever anyone tells of this example. One would think it possible to imagine, by a vast conceptual leap, some other example: a moth in Omaha, perhaps, or a starling in Sheboygan. Whatever winged creature is responsible, the point is that any small change in a chaotic system can, and typically does, have large and amplifying effects. Thus this sensitivity implies that the detailed initial condition—how fast, at what angle, and precisely how the starling flapped its wings—would have to be known to infinite precision to predict the result. But both practical and quantum considerations preclude such a possibility. Thus the familiar conclusion: for chaotic systems, we cannot predict long-term behavior. Note again that failure to predict does not mean failure to understand or to explain. Indeed, if we were confident we knew the equations governing a chaotic system, we would be confident we understood its behavior, including our incapacity to predict in detail its long-term behavior.

If we will often, in general and in principle, be precluded from making detailed predictions with the final theory, then what can we hope for? I once listened with considerable interest to an interior decorator whose design sense was demonstrably better than my own. I learned a useful phrase: it's that kind of thing. Now here is a phrase that has quite cosmic usefulness, for even in our incapacity to predict details, we can still have every hope of predicting kinds of things. The hope here is to characterize classes of properties of systems that, in senses to be made more precise later, are typical or generic and do not depend on the details. For example, when water freezes, one does not know where every water molecule is, but a lot can be said about your typical lump of ice. It has a characteristic temperature, color, and hardness—"robust" or "generic" features that do not depend on the details of its construction. And so it might be with complex systems such as organisms and economies. Not knowing the details we nevertheless can build theories that seek to explain the generic properties.

Advances in theoretical science have often been based on finding useful compact descriptions of a phenomenon of interest. The reduced description does not capture all the features of the phenomenon, just those that are fundamentally relevant. A simple example is the pendulum of a grandfather clock, or, in fancier terms, a harmonic oscillator. The pendulum might be described in terms of composition, length, weight, color, engravings on its surface, distance from other objects, and so on. But to understand the fundamental property of periodic motion, length and mass are important; the rest are not. Statistical mechanics gives us our clearest example of the use of statistically averaged, hence typical and generic, properties as compact descriptors of a complex system. Temperature and pressure are averaged properties of a volume of gas at equilibrium that are typically insensitive to the detailed behaviors of individual gas molecules.

Statistical mechanics demonstrates that we can build theories about those properties of complex systems that are insensitive to the details. But the statistical mechanics of gases is relatively simple, for all the gas molecules obey the same Newtonian laws of motion, and we seek to understand the averaged collective motions of the gas molecules. Familiar statistical mechanics concerns simple random systems. Organisms are not simple random systems, but highly complex, heterogeneous systems that have evolved for almost 4 billion years. Discovering the existence of key biological properties of complex living systems that do not depend on all the details lies at the center of a hope to build a deep theory of biological order. If all properties of living systems depend on every detail of their structure and logic, if organisms are arbitrary widgets inside arbitrary contraptions all the way down, then the epistemological problems that confront us in attempting to understand the wonder of the biosphere will be vast. If, instead, core phenomena of the deepest importance do not depend on all the details, then we can hope to find beautiful and deep theories. For example, ontogeny, the development of a fertilized egg into an adult, is controlled by networks of genes and their products in each cell of the body. If this unfolding depends on every small detail of the network, then understanding the order in organisms would require knowing all those details. Instead, I shall give strong grounds in later chapters to think that much of the order seen in development arises almost without regard for how the networks of interacting genes are strung together. Such order is robust and emergent, a kind of collective crystallization of spontaneous structure. Here is order whose origin and character we can hope to explain independently of the details. Here is spontaneous order that selection then goes on to mold.

The search for such properties is emerging as a fundamental research strategy, one I shall make much use of in this book. In these cases, one hopes to explain, understand, and even predict the occurrence of these generic emergent properties; however, one gives up the dream of predicting the details. Examples that we shall explore include the origin of life as a collective emergent property of complex systems of chemicals, the development of the fertilized egg into the adult as an emergent property of complex networks of genes controlling one another's activities, and the behavior of coevolving species in ecosystems that generates small and large avalanches of extinction and speciation. In all these cases, the order that emerges depends on robust and typical properties of the systems, not on the details of structure and function. Under a vast range of different conditions, the order can barely help but express itself.

But how would such laws of emergent order, if they should someday be found, be reconciled with the random mutations and opportunistic selections of Darwinism? How can life be contingent, unpredictable, and accidental while obeying general laws? The same question arises when we think about history. Historians have differed, some eschewing any hope of finding general laws. I, surely no historian, shall nevertheless have suggestions to make. For the hope arises that viewed on the most general level, living systems—cells, organisms, economies, societies—may all exhibit lawlike properties, yet be graced with a lacework of historical filigree, those wonderful details that could easily have been otherwise, whose very unlikelihood elicits our awed admiration.

And so we return to the overarching question: Whence cometh all this bubbling activity, complexity, and chutzpah? It is our quest to understand the emergence of this ordered complexity around us, in the living forms we see, the ecosystems they construct, the social systems that abound from insects to primates, the wonder of economic systems that actually deliver us our daily bread and astonished Adam Smith into the conceptualization of his Invisible Hand. I am a doctor-biologist. I am hopeful enough to think I might possibly help in understanding the origin of life and its subsequent evolution. I am not a physicist. I am not brazen enough to presume to think about cosmic evolution. But I wonder: Whence cometh all this bubbling activity and complexity? Ultimately, it must be a natural expression of a universe that is not in equilibrium, where instead of the featureless homogeneity of a vessel of gas molecules, there are differences, potentials, that drive the formation of complexity. The flash of the Big Bang 15 billion years ago has yielded a universe said to be expanding, perhaps never to fall together in the Big Crunch. It is a *nonequilibrium* universe filled with too many hydrogen and helium atoms compared with the most stable atomic form, iron. It is a universe of galaxies and clusters of galaxies on many scales, where none might have formed at all. It is a universe of stunningly abundant free energy available for performing work. The life around us must somehow be the natural consequence of the coupling of that free energy to forms of matter. How? No one knows. But we shall hazard hypotheses along the way. Here is no mere scientific search. Here is a mystical longing, a sacred core first sought around that small campfire sometime in the past 3 million years. This way lies the search for our roots. If we are, in ways we do not yet see, natural expressions of matter and energy coupled together in nonequilibrium systems, if life in its abundance were bound to arise, not as an incalculably improbable accident, but as an expected fulfillment of the natural order, then we truly are at home in the universe.

Physicists, chemists, and biologists are familiar with two major forms by which order arises. The first involves what are known as low-energy equilibrium systems. A familiar example is a ball in a bowl that rolls to the bottom, wobbles a bit, and stops. The ball stops at a position that minimizes its potential energy. Its kinetic energy of motion, acquired because of gravity, has been dissipated into heat by friction. Once the ball is at equilibrium, located at the bottom of the bowl, no further input of energy is needed to maintain that spatial order. In biology, similar examples abound. Thus viruses are complex molecular systems of DNA or RNA molecular strands that form a core around which a variety of proteins assemble to form tail fibers, head structures, and other features. In an appropriate aqueous environment, the viral particle will self-assemble from its molecular DNA or RNA and protein constituents, seeking its state of lowest energy, like the ball in the bowl. Once the virus is formed, no further input of energy is required to maintain it.

The second means by which order arises requires a constant source of mass or energy or both to sustain the ordered structure. Unlike the ball in the bowl, such systems are nonequilibrium structures. A whirlpool in a bathtub is a familiar example. Once formed, the nonequilibrium swirl can be stable for long periods if water is continuously added to the tub and the drain is left open. One of the most startling examples of such a sustained nonequilibrium structure is the Great Red Spot on Jupiter, which appears to be a whirlpool in the upper atmosphere of that enormous planet. The Great Red Spot vortex, essentially a storm system, has been present for at least several centuries. Thus the lifetime of the Great Red Spot is far longer than the average time any single gas molecule has lingered within it. It is a stable organization of matter and energy through which both matter and energy flow. The similarity to a human organism, whose molecular constituents change many times during a lifetime, is intriguing. One can have a remarkably complex discussion about whether the Great Red Spot might be considered to be living—and if not, why not. After all, the Great Red Spot in some sense persists and adapts to its environment, shedding baby vortices as it does so.

Nonequilibrium ordered systems like the Great Red Spot are sustained by the persistent dissipation of matter and energy, and so were named dissipative structures by the Nobel laureate Ilya Prigogine some decades ago. These systems have received enormous attention. In part, the interest lies in their contrast to equilibrium thermodynamic systems, where equilibrium is associated with collapse to the most probable, least ordered states. In dissipative systems, the flux of matter and energy through the system is a driving force generating order. In part, the interest lies in the awareness that free-living systems are dissipative structures, complex metabolic whirlpools. Here I am being careful to make a distinction between free-living systems and viruses. Viruses are not free-living entities; rather, they are parasites that must invade cells to reproduce. All known free-living systems are composed of cells, from bacteria to bottleflies. Cells are not low-energy structures. Cells hum along as complex chemical systems that persistently metabolize food molecules to maintain their internal structure and to reproduce. Hence cells are nonequilibrium dissipative structures. Interestingly, some simple cells such as spore forms, which presumably *are* low-energy structures, can enter nonmetabolizing quiescent states. For most cells, however, equilibrium corresponds to death.

Since all free-living systems are nonequilibrium systems—indeed, since the biosphere itself is a nonequilibrium system driven by the flux of solar radiation—it would be of the deepest importance were it possible to establish general laws predicting the behavior of all nonequilibrium systems. Unfortunately, efforts to find such laws have not yet met with success. Some believe that they may never be discovered. The failure may not be a lack of cleverness on our part, but a consequence of a well-founded area of mathematics called the theory of computation. This beautiful theory is concerned with what are called effectively computable algorithms. Algorithms are a set of procedures to generate the answer to a problem. An example is the algorithm to find the solution to a quadratic equation, which most of us were taught while learning algebra. Not only was I taught the algorithm, but our entire algebra class was invited to tattoo it on our tummies so that we could solve quadratic equations by rote. In short, algorithms can be carried out by any careful dummy. Computers are just such dummies, and familiar computer programs are just algorithms.

The theory of computation is replete with deep theorems. Among the most beautiful are those showing that, in most cases by far, there exists no shorter means to predict what an algorithm will do than to simply execute it, observing the succession of actions and states as they unfold. The algorithm itself is its own shortest description. It is, in the jargon of the field, incompressible.

The next step in the argument that no general laws could predict the detailed behavior of all nonequilibrium systems is simple. Real computers, made of real materials and plugged into a wall socket, are what Alan Turing called universal computational systems. He showed that with an infinitely long memory tape, a universal computer could carry out any algorithm at all. A physical computer qualifies as a nonequilibrium system; supplied with a constant source of energy, it is able to carry out calculations by using energy to maneuver electronic bits in silicon chips in various patterns. But the theory of computation tells us that such a device might be behaving in a way that is *its own shortest description*. The shortest way to predict what this real physical system will do is just to watch it. But the very purpose of a theory is to provide just such a shorter, compressed description—Kepler's laws instead of a catalog of every position of every planet at every moment in time. Since such a physical computer is a real nonequilibrium system, we could not have a general theory predicting the detailed behavior of all possible nonequilibrium systems. But cells, ecosystems, and economic systems are also real nonequilibrium systems. It is conceivable that these, too, behave in ways that are their own shortest descriptions.

In considering whether there can be laws of life, many biologists would answer with a firm no. Darwin has properly taught us of descent with modification. Modern biology sees itself as a deeply historical science. Shared features among organisms—the famous genetic code, the spinal column of the vertebrates—are seen not as expressions of underlying law, but as contingent useful accidents passed down through progeny as useful widgets, found and frozen thereafter into that descendant branch of life. It is by no means obvious that biology will yield laws beyond descent with modification. But I believe that such laws can be found.

We wish to understand the order around us in the biosphere, and we see that that order may reflect both low-energy equilibrium forms (the ball in the bowl, the virus) and nonequilibrium, dissipative structures, the living whirlpools that maintain order by importing and exporting matter and energy. Yet we have now encountered at least three difficulties that stand in our way. First, quantum theory precludes detailed prediction of molecular phenomena. Whatever the final theory, this world has seen too many throws of the quantum dice to predict its detailed state. Second, even were classical determinism to hold, the theory of chaos shows us that very small changes in initial conditions can lead to profound changes of behavior in a chaotic system. As a matter of practicality, we may typically be unable to know initial conditions with sufficient precision to predict detailed behavior. Finally, the theory of computation seems to imply that nonequilibrium systems can be thought of as computers carrying out algorithms. For vast classes of such algorithms, no compact, lawlike description of their behavior can be obtained.

If the origin and evolution of life is like an incompressible computer algorithm, then, in principle, we can have no compact theory that predicts all the details of the unfolding. We must instead simply stand back and watch the pageant. I suspect that this intuition may prove correct. I suspect that evolution itself is deeply like an incompressible algorithm. If we demand to know its details, we must watch in awed wonder and count and recount the myriad rivulets of branching life and the multitudes of its molecular and morphological details.

And yet, even if it is true that evolution is such an incompressible process, it does not follow that we may not find deep and beautiful laws governing that unpredictable flow. For we are not precluded from the possibility that many features of organisms and their evolution are profoundly robust and insensitive to details. If, as I believe, many such robust properties exist, then deep and beautiful laws may govern the emergence of life and the population of the biosphere. After all, what we are after here is not necessarily detailed prediction, but explanation. We can never hope to predict the exact branchings of the tree of life, but we can uncover powerful laws that predict and explain their general shape. I hope for such laws. I even dare to hope that we can begin to sketch some of them now. For want of a better general phrase, I call these efforts a search for a theory of emergence.

Order for Free

The vast mystery of biology is that life should have emerged at all, that the order we see should have come to pass. A theory of emergence would account for the creation of the stunning order out our windows as a natural expression of some underlying laws. It would tell us if we are at home in the universe, expected in it, rather than present despite overwhelming odds.

Some words or phrases are evocative, even provocative. So it is with the word *emergent*. Commonly, we express this idea with the sentence, The whole is greater than the sum of its parts. The sentence is provocative, for what extra can be in the whole that is not in the parts? I believe that life itself is an emergent phenomenon, but I mean nothing mystical by this. In Chapters 2 and 3, I shall be at pains to give good reasons to believe that sufficiently complex mixes of chemicals can spontaneously crystallize into systems with the ability to collectively catalyze the network of chemical reactions by which the molecules themselves are formed. Such collectively autocatalytic sets sustain themselves and reproduce. This is no less than what we call a living metabolism, the tangle of chemical reactions that power every one of our cells. Life, in this view, is an emergent phenomenon arising as the molecular diversity of a prebiotic chemical system increases beyond a threshold of complexity. If true, then life is not located in the property of any single molecule—in the details—but is a collective property of systems of interacting molecules. Life, in this view, emerged whole and has always remained whole. Life, in this view, is not to be located in its parts, but in the collective emergent properties of the whole they create. Although life as an emergent phenomenon may be profound, its fundamental holism and emergence are not at all mysterious. A set of molecules either does or does not have the property that it is able to catalyze its own formation and reproduction from some simple food molecules. No vital force or extra substance is present in the emergent, self-reproducing whole. But the collective system does possess a stunning property not possessed by any of its parts. It is able to reproduce itself and to evolve. The collective system is alive. Its parts are just chemicals.

One of the most awesome aspects of biological order is ontogeny, the development of an adult organism. In humans, this process starts with a single cell, the fertilized egg, or zygote. The zygote undergoes about 50 cell divisions to create about 1 quadrillion cells that form the newborn infant. At the same time, the single cell type of the zygote differentiates to form the roughly 260 cell types of the adult—liver parenchymal cells, nerve cells, red blood cells, muscle cells, and so forth. The genetic instructions controlling development lie in the DNA within the nucleus of each cell. This genetic system harbors about 100,000 different genes, each encoding a different protein. Remarkably, the set of genes in all cell types is virtually identical. Cells differ because different subsets of genes are active within them, producing various enzymes and other proteins. Red blood cells have hemoglobin, muscle cells abound in the actin and myosin that form muscle fibers, and so forth. The magic of ontogeny lies in the fact that genes and their RNA and protein products form a complex network, switching one another on and off in a wondrously precise manner.

We can think of this genomic system as a complex chemical computer, but this computer differs from familiar serial-processing computers, which carry out one action at a time. In the genomic computer system, many genes and their products are active at the same time; hence the system is a parallel-processing chemical computer of some kind. The different cell types of the developing embryo and its trajectory of development are, in some sense, expressions of the behavior of this complex genomic network. The network within each cell of any contemporary organism is the result of at least 1 billion years of evolution. Most biologists, heritors of the Darwinian tradition, suppose that the order of ontogeny is due to the grinding away of a molecular Rube Goldberg machine, slapped together piece by piece by evolution. I present a countering thesis: most of the beautiful order seen in ontogeny is spontaneous, a natural expression of the stunning self-organization that abounds in very complex regulatory networks. We appear to have been profoundly wrong. Order, vast and generative, arises naturally.

The emergent order seen in genomic networks foretells a conceptual struggle, perhaps even a conceptual revolution, in evolutionary theory. In this book, I propose that much of the order in organisms may not be the result of selection at all, but of the spontaneous order of self-organized systems. Order, vast and generative, not fought for against the entropic tides but freely available, undergirds all subsequent biological evolution. The order of organisms is natural, not merely the unexpected triumph of natural selection. For example, I shall later give strong grounds to think that the homeostatic stability of cells (the biological inertia that keeps a liver cell, say, from turning into a muscle cell), the number of cell types in an organism compared with the number of its genes, and other features are not chance results of Darwinian selection but part of the order for free afforded by the self-organization in genomic regulatory networks. If this idea is true, then we must rethink evolutionary theory, for the sources of order in the biosphere will now include both selection *and* self-organization.

This is a massive and difficult theme. We are just beginning to embrace it. In this new view of life, organisms are not merely tinkered-together contraptions, bricolage, in Jacob's phrase. Evolution is not merely "chance caught on the wing," in Monod's evocative image. The history of life captures the natural order, on which selection is privileged to act. If this idea is true, many features of organisms are not merely historical accidents, but also reflections of the profound order that evolution has further molded. If true, we are at home in the universe in ways not imagined since Darwin stood natural theology on its head with his blind watchmaker.

Yet more is presaged by self-organization. I said we must encompass the roles of both self-organization *and* Darwinian selection in evolution. But these sources of order may meld in complex ways that we hardly begin to understand. No theory in physics, chemistry, biology, or elsewhere has yet brokered this marriage. We must think anew. Among the progeny of this mating of self-organization and selection may be new universal laws.

It is perhaps astonishing, perhaps hopeful and wonderful, that we might even now begin to frame possible universal laws governing this proposed union. For what can the teeming molecules that hustled themselves into self-reproducing metabolisms, the cells coordinating their behaviors to form multicelled organisms, the ecosystems, and even economic and political systems have in common? The wonderful possibility, to be held as a working hypothesis, bold but fragile, is that on many fronts, life evolves toward a regime that is poised between order and chaos. The evocative phrase that points to this working hypothesis is this: life exists at the edge of chaos. Borrowing a metaphor from physics, life may exist near a kind of phase transition. Water exists in three phases: solid ice, liquid water, and gaseous steam. It now begins to appear that similar ideas might apply to complex adapting systems. For example, we will see that the genomic networks that control development from zygote to adult can exist in three major regimes: a frozen ordered regime, a gaseous chaotic regime, and a kind of liquid regime located in the region between order and chaos. It is a lovely hypothesis, with considerable supporting data, that genomic systems lie in the ordered regime near the phase transition to chaos. Were such systems too deeply into the frozen ordered regime, they would be too rigid to coordinate the complex sequences of genetic activities necessary for development. Were they too far into the gaseous chaotic regime, they would not be orderly enough. Networks in the regime near the edge of chaos—this compromise between order and surprise—appear best able to coordinate complex activities and best able to evolve as well. It is a very attractive hypothesis that natural selection achieves genetic regulatory networks that lie near the edge of chaos. Much of this book is bent on exploring this theme.

Evolution is a story of organisms adapting by genetic changes, seeking to improve their fitness. Biologists have long harbored images of fitness landscapes, where the peaks represent high fitness, and populations wander under the drives of mutation, selection, and random drift across the landscape seeking peaks, but perhaps never achieving them.

The idea of fitness peaks applies at many levels. For example, it can refer to the capacity of a protein molecule to catalyze a given chemical reaction. Then peaks of the landscape correspond to enzymes that are better catalysts for this reaction than all their neighboring proteins—those in the foothills and, worst of all, those in the valleys. Fitness peaks can also refer to the fitness of whole organisms. In that more complex case, an organism with a given set of traits is fitter—higher on the landscape—than all its near variants if, roughly speaking, it is more likely to have offspring.

We will find in this book that whether we are talking about organisms or economies, surprisingly general laws govern adaptive processes on multipeaked fitness landscapes. These general laws may account for phenomena ranging from the burst of the Cambrian explosion in biological evolution, where taxa fill in from the top down, to technological evolution, where striking variations arise early and dwindle to minor improvements. The edge-of-chaos theme also arises as a potential general law. In scaling the top of the fitness peaks, adapting populations that are too methodical and timid in their explorations are likely to get stuck in the foothills, thinking they have reached as high as they can go; but a search that is too wide ranging is also likely to fail. The best exploration of an evolutionary space occurs at a kind of phase transition between order and disorder, when populations begin to melt off the local peaks they have become fixated on and flow along ridges toward distant regions of higher fitness.

The edge-of-chaos image arises in coevolution as well, for as we evolve, so do our competitors; to remain fit, we must adapt to their adaptations. In coevolving systems, each partner clambers up its fitness landscape toward fitness peaks, even as that landscape is constantly deformed by the adaptive moves of its coevolutionary partners. Strikingly, such coevolving systems also behave in an ordered regime, a chaotic regime, and a transition regime. It is almost spooky that such systems seem to coevolve to the regime at the edge of chaos. As if by an invisible hand, each adapting species acts according to its own selfish advantage, yet the entire system appears magically to evolve to a poised state where, on average, each does as best as can be expected. Yet, as in many of the dynamical systems we will study in this book, each is eventually driven to extinction, despite its own best efforts, by the collective behavior of the system as a whole.

As we shall see, technological evolution may be governed by laws similar to those governing prebiotic chemical evolution and adaptive coevolution. The origin of life at a threshold of chemical diversity follows the same logic as a theory of economic takeoff at a threshold of diversity of goods and services. Above that critical diversity, new species of molecules, or goods and services, afford niches for yet further new species, which are awakened into existence in an explosion of possibilities. Like coevolutionary systems, economic systems link the selfish activities of more or less myopic agents. Adaptive moves in biological evolution and technological evolution drive avalanches of speciation and extinction. In both cases, as if by an invisible hand, the system may tune itself to the poised edge of chaos where all players fare as well as possible, but ultimately exit the stage.

The edge of chaos may even provide a deep new understanding of the logic of democracy. We have enshrined democracy as our secular religion; we argue its moral and rational foundations, and base our lives on it. We hope that our heritage of democracy will spill out its abundance of freedom over the globe. And in the following chapters we will find surprising new grounds for the secular wisdom of democracy in its capacity to solve extremely hard problems characterized by intertwining webs of conflicting interests. People organize into communities, each of which acts for its own benefit, jockeying to seek compromises among conflicting interests. This seemingly haphazard process also shows an ordered regime where poor compromises are found quickly, a chaotic regime where no compromise is ever settled on, and a phase transition where compromises are achieved, but not quickly. The best compromises appear to occur at the phase transition between order and chaos. Thus we will see hints of an apologia for a pluralistic society as the natural design for adaptive compromise. Democracy may be far and away the best process to solve the complex problems of a complex evolving society, to find the peaks on the coevolutionary landscape where, on average, all have a chance to prosper.

Wisdom, Not Power

I suggest in the ensuing chapters how life may have formed as a natural consequence of physics and chemistry, how the molecular complexity of the biosphere burgeoned along a boundary between order and chaos, how the order of ontogeny may be natural, and how general laws about the edge of chaos may govern coevolving communities of species, of technologies, and even of ideologies.

This poised edge of chaos is a remarkable place. It is a close cousin of recent remarkable findings in a theory physicists Per Bak, Chao Tang, and Kurt Wiesenfeld called self-organized criticality. The central image here is of a sandpile on a table onto which sand is added at a constant slow rate. Eventually, the sand piles up and avalanches begin. What one finds are lots of small avalanches and few large ones. If the size of the avalanche is plotted on the familiar x-axis of a Cartesian coordinate system, and the number of avalanches at that size are plotted on the y-axis, a curve is obtained. The result is a relationship called a power law. The particular shape of this curve, to which we shall return in later chapters, has the stunning implication that the same-sized grain of sand can unleash small or large avalanches. Although we can say that in general there will be more tiny avalanches and only a few big landslides (that is the nature of a power-law distribution), there is no way to tell whether a particular one will be insignificant or catastrophic.

Sandpiles, self-organized criticality, and the edge of chaos. If I am right, the very nature of coevolution is to attain this edge of chaos, a web of compromises where each species prospers as well as possible but where none can be sure if its best next step will set off a trickle or a landslide. In this precarious world, avalanches, small and large, sweep the system relentlessly. One's own footsteps shed small and large avalanches, which sweep up or by the other hikers on the slopes below. One may even be carried off in the avalanche started by his or her own footsteps. This image may capture the essential features of the new theory of emergence we seek. At this poised state between order and chaos, the players cannot foretell the unfolding consequences of their actions. While there is law in the distribution of avalanche sizes that arise in the poised state, there is unpredictability in each individual case. If one can never know if the next footstep is the one that will unleash the landslide of the century, then it pays to tread carefully.

In such a poised world, we must give up the pretense of long-term prediction. We cannot know the true consequences of our own best actions. All we players can do is be locally wise, not globally wise. All we can do, all anyone can do, is hitch up our pants, put on our galoshes, and get on with it the best we can. Only God has the wisdom to understand the final law, the throws of the quantum dice. Only God can foretell the future. We, myopic after 3.45 billion years of design, cannot. We, with all the others, cannot foretell the avalanches and their intertwinings that we jointly generate. We can do only our local, level best. We can get on with it.

Since the time of Bacon our Western tradition has regarded knowledge as power. But as the scale of our activities in space and time has increased, we are being driven to understand the limited scope of our understanding and even our potential understanding. If we find general laws, and if those laws entail that the biosphere and all within it co-evolve to some analogue of the sandpile, poised on the edge of chaos, it would be wise to be wise. We enter a new millennium. It is best to do so with gentle reverence for the ever-changing and unpredictable places in the sun that we craft ever anew for one another. We are all at home in the universe, poised to sanctify by our best, brief, only stay.

Chapter 2

The Origins
of Life

A nyone who tells you that he or she knows how life started on the sere earth some 3.45 billion years ago is a fool or a knave. Nobody knows. Indeed, we may never recover the actual historical sequence of molecular events that led to the first self-reproducing, evolving molecular systems to flower forth more than 3 million millennia ago. But if the historical pathway should forever remain hidden, we can still develop bodies of theory and experiment to show how life might realistically have crystallized, rooted, then covered our globe. Yet the caveat: nobody knows.

First was the word, then the cleaving of darkness from light. By the third day, life-forms were crafted: fish, fowl, and more. Adam and Eve awakened on the sixth day. Not so wrong, after all, is this mythos in which life was believed to arise so rapidly. In reality, life sprang from the molten earth's womb about as soon as the infall of meteoric masses forming the protoearth slowed substantially and the surface cooled enough to support liquid water, an arena in which chemicals could combine to produce metabolisms. The earth is about 4 billion years old. No one knows what the earliest self-reproducing molecular systems looked like. But by 3.45 billion years ago, archaic forms of cells fingered some clay or rock surfaces, were buried, and left their traces for our later questions. I am no expert on such ancient fossils, but was delighted in Chapter 1 to share the beautiful work of William Schopf and his colleagues around the world. Figures 1.1 and 1.2 show some of the earliest fossil cells.

What prodigious progress these early cells exhibit! One guesses from their morphology that, like contemporary cells, their membrane was a bilipid layer—a kind of double-layered soap bubble made of fatty lipid molecules—enclosing a network of molecules that had the power to

31

sustain itself and reproduce. But how could such self-reproducing assemblages of molecular effort have coagulated from the primeval cloud of hydrogen and larger atoms and molecules, which itself coagulated from a dust cloud into the early earth? We stand, with *Homo habilis,* in need of a creation myth. This time, armed with the power of late-twentieth-century science, perhaps we can stumble on the truth.

Theories of Life

The question of the origin of life has undergone major transformations over the past centuries. This is not so surprising. In the Western tradition a thousand years ago, most who thought of the problem were persuaded that life formed spontaneously from nonlife. After all, maggots seemed to appear from nothing, in fruits and rotting wood, and full-blown adult insects hastened forth from their metamorphic chrysalis tents. Life sprang unbidden from moldering places, damp with promise. Spontaneous generation was but another miracle of the daily sort that God's hand shaped—quotidian, common, sustaining.

Theories of the origin of life began to appear in their modern forms only with the brilliant experiments of Louis Pasteur over a century ago. How could one mind have done so much? A prize was offered for the most telling experiments concerning the theory of spontaneous generation. The growth of bacterial populations had been demonstrated within what were believed to be initially sterile solutions. Pasteur rightly suspected that the source of the bacteria was the air itself, for the flasks used by his predecessors were open and crafted in such a way that it would be easy for bacteria to drift into the broth. Pasteur set about creating flasks with swan-necked, S-shaped openings. He hoped that any bacteria that entered from the outside would be trapped before they could reach the broth. Simple, elegant experiments always delight us the most. Pasteur found no growth of bacteria in his sterile broth. Life, he concluded, came from life.

But if life comes only from life, then whence life in the first place? With Pasteur, the problem of origins looms suddenly vast, profound, mysterious, perhaps unstatable, perhaps beyond science itself. Alchemy had led to chemistry, which led to the analysis of inorganic atoms and molecules: lead, copper salts, gold, oxygen, hydrogen. But organisms harbored molecules not found in nonliving materials. Organisms harbored organic molecules. The difference between living and nonliving was, for some time, supposed to reside in these different kinds of molecules. No bridge could close the gap. Then, midway through the nineteenth century, Emil Fischer synthesized urea, clearly an organic compound, from inorganic chemicals. Life was made of the same stuff as nonlife. Fischer's result carried the implication that common physical and chemical principles might govern both living and nonliving matter. His achievement remains a major step in the reduction of biology to chemistry and physics. In a way, the believers in spontaneous generation were right after all: life did come from nonlife, though this conjuring act was far more complex than they ever could have supposed.

But the reductionist thesis that life is based on the same principles as nonlife was not to be so readily accepted. For even if it is granted that life is cut from the same cloth as nonlife, it does not follow that the cloth by itself suffices. The clothes need not, in truth, quite make the man. The French philosopher Henri Bergson proposed an answer to this wonderful mystery that persuaded many for decades: élan vital. Like any other fine French perfume, without which flesh was, well, merely flesh, élan vital was said to be an insubstantial essence that permeated and animated the inorganic molecules of cells and brought them to life. Was this really so silly? It is easy to be smug until our own cherished certitudes crumble. After all, frog muscles had recently been shown to exhibit animal magnetism—now more properly understood as the electric potential changes that propagate along nerve and muscle fibers—and the magnetic field of James Clerk Maxwell was itself insubstantial yet able to move matter held in its sway. If an insubstantial magnetic field could move solid matter, why could not an insubstantial élan vital animate the inanimate?

Bergson was not the only thoughtful person to advance such vitalist ideas. Hans Dreisch, a brilliant experimentalist, came to much the same conclusions. Dreisch had performed experiments on the two-celled frog embryo. Like most other embryos, the fertilized cell, or zygote, divides again and again, creating 2, 4, 8, 16 cells, and on and on until a creature is born. Dreisch looped a blond hair of a child around the embryo, pinching its two cells free from each other. To his utter astonishment, each cell developed into a fully normal frog! Even single cells isolated from later embryos, consisting of four and eight cells, could give rise to entire adult frogs.

Dreisch was no fool. He realized he had a hell of a puzzle on his hands. Nothing obvious in the entire Newtonian tradition of physical and chemical science appeared to offer even a glimmer of hope to explain such an astonishing result. It would have been acceptable if each part of the embryo gave rise to one part of an adult; indeed, this occurs in the embryos of many species and is called mosaic development. Mosaic development might be understood using the arguments espoused by a group called the preformationists. The egg supposedly housed a homunculus, a tiny version of the adult, each part of which somehow expanded into the corresponding part of the adult. Thus deletion of half the egg—one of the two daughter cells of the zygote—would be expected to delete half the homunculus. The remaining half-egg or single cell should yield half a frog. But that is not what happened. Even if it had, the preformationists would still have been left with the overwhelming problem of explaining how the newly formed adult gave birth to a child that grew to an adult and gave birth to another child, and so on down the genealogical line. The preformationists suggested that the problem could be solved by supposing that, inside the egg, homunculi were nested inside homunculi like Chinese dolls—all the way back to creation. If life were to persist forever, of course, an infinity of such nested homunculi would be needed. Here, I confess, my willingness to sympathize with out-of-date ideas fades away. Theories, even those that prove incorrect, can have elegance and beauty, or be utterly ad hoc. A theory requiring an infinite series of ever smaller homunculi is too ad hoc to be true.

Dreisch had made an important discovery. If each cell in the two- or four- or eight-celled embryo could give rise to an entire adult, then the information had to come from somewhere. Somehow, the order was emergent; each part could give rise to the whole. But where did the information come from within each part? Dreisch turned to what were called entelechies, nonmaterial sources of order that invested the embryo and its mere stuff, and somehow led to the capacity of each part to give rise, magically, to the whole.

The origin-of-life problem rested quietly from the late nineteenth century for some 50 or more years, held by most thinkers to be either unapproachable scientifically or, at best, so premature that any efforts were hopeless. In the middle of the twentieth century, attention turned to the nature of the primitive earthly atmosphere that gave rise to the chemicals of life. Good evidence, now subject to some doubt, suggested that the early atmosphere was rich in molecular species such as hydrogen, methane, and carbon dioxide. Almost no oxygen was present. Further, it was supposed, simple organic molecules in the atmosphere, along with other more complex ones, would be expected to dissolve slowly in the newly formed oceans, creating a prebiotic soup. From this soup, it was hoped, life would somehow form spontaneously.

This hypothesis continues to have many adherents, though it suffers from considerable difficulties. Chief among them is the fact that the soup would be extremely dilute. The rate of chemical reactions depends on how rapidly the reacting molecular species encounter one another—

and that depends on how high their concentrations are. If the concentration of each is low, the chance that they will collide is very much lower. In a dilute prebiotic soup, reactions would be very slow indeed. A wonderful cartoon I recently saw captures this. It was entitled "The Origin of Life." Dateline 3.874 billion years ago. Two amino acids drift close together at the base of a bleak rocky cliff; three seconds later, the two amino acids drift apart. About 4.12 million years later, two amino acids drift close to each other at the base of a primeval cliff. . . . Well, Rome wasn't built in a day. Could life have crystallized in such a dilute medium—even if we waited as long as the age of the universe? We return in a moment to unhappy calculations, calculations that I find faulty but fun, which suggest that life could not have crystallized by chance in billions of times the lifetime of the universe. How unfortunate, for here I sit writing for you to read. Something must be wrong somewhere.

Alexander Oparin, a Russian biophysicist, writing from the hellhole of Stalinist life, proposed a plausible way to confront the problem afforded by the dilute soup. When glycerine is mixed with other molecules, it forms gel-like structures called coacervates. The coacervate is able to concentrate organic molecules inside itself and exchange them across its boundary. In short, coacervates are like primitive cells, which separate the molecular activities within from the dilute aqueous soup. If these tiny compartments had developed in the primeval waters, they might have concentrated the proper chemicals needed to form metabolisms.

If Oparin opened the door to understanding how protocells might have formed, it remained entirely unclear where their contents—the small organic molecules whose traffic is metabolism—came from. In addition to simple molecules are various polymers, long molecular chains of nearly identical building blocks. Proteins, of which muscles, enzymes, and the scaffolding of cells are made, consist of chains of 20 kinds of amino acids. This linear primary structure then folds up into a more or less compact three-dimensional structure. DNA and RNA are composed of chains of four nucleotide building blocks: adenine, cytosine, guanine, and thymine in DNA, with uracil substituting for thymine in RNA. Without these molecules, the very stuff of life, Oparin's coacervates would be no more than empty shells. Where, then, might these building blocks have come from?

In 1952, Stanley Miller, then a young graduate student in the laboratory of a famous chemist named Harold Urey, tried a dream of a crazy idea. He filled a flask with the gases—methane, carbon dioxide, and so forth—that were generally assumed to have been present in the atmosphere of the primitive earth. He showered the flask with sparks, mimicking lightning as a source of energy. He waited, hoping that he might be peering into a homemade Garden of Eden. Several days later, he was rewarded with evidence of molecular creativity: brown gunk clung to the sides and bottom of his flask. Upon analysis, this tarry material proved to contain a rich variety of amino acids. Miller had performed the first prebiotic chemistry experiment. He had discovered plausible means whereby the building blocks of proteins might have been formed on the early earth. He received his Ph.D. and has been a leader in the field of prebiotic chemistry ever since.

Similar experiments have shown that it is possible (though with much greater difficulty) to form the nucleotide building blocks of DNA, RNA, and fatty molecules and hence, through them, the structural material for cellular membranes. Many other small molecular components of organisms have been synthesized abiogenically.

But substantial puzzles remain. Robert Shapiro notes in his book *Origins: A Skeptic's Guide to the Creation of Life on Earth* that even though scientists can show that it is possible to synthesize the various ingredients of life, it is not easy to get them to cohere into a single story. One group of scientists discovers that molecule A can be formed from molecules B and C in a very low yield under a certain set of conditions. Then, having shown that it is possible to make A, another group starts with a high concentration of the molecule and shows that by adding D one can form E—again in a very low yield and under quite different conditions. Then another group shows that E, in high concentration, can form F under still different conditions. But how, without supervision, did all the building blocks come together at high enough concentrations in one place and at one time to get a metabolism going? Too many scene changes in this theater, argues Shapiro, with no stage manager.

The discovery of the molecular structure of genes, the famous DNA double helix, is the final event that underlies the resurging interest in the origin of life. Before James Watson and Francis Crick's famous paper in 1953, it was a matter of deep debate among biologists and biochemists whether the genetic material would prove to be protein or DNA. Those favoring proteins as the fundamental genetic material had much to say in favor of their hypothesis: most notably, almost all enzymes are proteins. Enzymes, of course, are the major class of biological catalysts, molecules that bind to substrates and speed up the rate of the reactions necessary to make a metabolism. In addition, many of the structural molecules in cells are proteins. A familiar example is hemoglobin, found in red blood cells and important in binding and transporting oxygen from the lungs to the tissues. Since proteins are ubiquitous and are the workhorses of the body's cellular scaffolding and metabolic flux, it was not unreasonable to suppose that these complex polymers of amino acids were also the carriers of genetic information.

An intellectual lineage beginning with Mendel, however, pointed to the chromosomes found in each cell as the carriers of genetic information. Most readers will be familiar with Mendel's beautiful genetic experiments on sweet peas in the 1870s. Atomism was the intellectual order of the day, for the burgeoning science of chemistry had found strong reasons to suppose that chemical reactions formed molecules out of simple whole-number ratios of their atomic building blocks. Water is precisely H_2O—two hydrogens (never two and a half) and one oxygen.

If atoms underlie chemistry, might there not be atoms of heredity? Children look something like each parent. Suppose this were caused by atoms of heredity, some from the mother and some from the father. But parents have parents, back for vast numbers of generations. If all the atoms of heredity were passed from each parent to each offspring, an enormous number of them would accumulate. To prevent this, on average, each offspring should receive only half the heredity atoms from each parent. The simplest hypothesis is that each offspring receives exactly one atom of heredity per trait from each parent. The two atoms would determine the trait—blue eyes versus brown eyes, for example—and would be passed on in turn to the next generation.

The rediscovery of Mendel's laws in 1902—nobody paid any attention to his initial discovery—is one of biology's heart-warming stories. Chromosomes, so named because of the colored stain that allowed them to show up under a microscope, had been identified in the nuclei of plant and animal cells. At cell division, or mitosis, the nucleus also divides. Each chromosome within the nucleus is first duplicated, and then one copy passes to each daughter nucleus, hence to each daughter cell. But even more impressive is the cellular process called meiosis, by which sperm and egg are formed. In meiosis, the number of chromosomes that reach the sperm or egg is exactly half that which is in the other cells of the body. Only when egg combines with sperm to produce a zygote is the full genetic inheritance restored. Each regular cell in the body, called a somatic cell, has pairs of chromosomes: one from the father, one from the mother. Further work showed that when egg or sperm cells are formed, either the maternal or the paternal chromosome of each pair is chosen at random and passed along. Since Mendel's laws demand that each parent pass a randomly chosen half of its genetic instructions to the offspring, the conclusion was hardly escapable: the chromosomes must be the carriers of genetic information. The flowering of experimental genetics had, by the 1940s, afforded overwhelming confirmation to this belief.

But the chromosomes are made primarily of a complex polymer called DNA, or dioxyribonucleic acid. Thus it seemed likely that genes, the new name for atoms of heredity, might be made from DNA. A famous experiment by microbiologist Oswald Avery settled the issue. Avery coaxed bacteria to take up some pure DNA derived from other bacteria. The recipient bacteria then exhibited some traits of the donor bacteria, and the new trait was stably inherited when the bacteria divided. DNA could carry heritable genetic information.

The race was on to discover what it is about DNA that allows it to encode this information. The story of the double helix, with its complementary strands, is famous. DNA, heralded as the master molecule of life—a view with which I both agree and profoundly disagree—proved to be a double helix of four nucleotide bases: adenine (A), guanine (G), cytosine (C), and thymine (T). As most readers know, the magic lies in the specific base pairing: A bonds specifically to T; C bonds specifically to G. The genetic information is carried in the sequence of bases along one or the other strand of the double helix. Triplets of bases—AAA, GCA, and so forth—specify each amino acid. Thus the sequence of bases can be translated by the cell into a specific sequence of amino acids to form a precise protein.

It is hard not to marvel at how the double-helix structure of DNA immediately suggested how the molecule might replicate. Each strand specifies the nucleotide sequence of the complementary strand by the precise A–T and C–G base pairing. Knowing the sequence along one strand, call it Watson, tells one what the sequence along the other strand, Crick, must be.

If DNA is a double helix in which each strand is the complement of the other, if the sequence of bases along Watson specifies the sequence of bases along Crick and vice versa, then the DNA double helix might be a molecule able to replicate itself spontaneously. DNA, in short, becomes a candidate for the first living molecule. The very molecule that is hailed as the master molecule of present life, the carrier of the genetic program by which the organism is computed from the fertilized egg, the very same magical molecule might have been the first self-reproducing molecule at the dawn of life. It would multiply, eventually chancing on the recipe to make proteins to clothe itself and to speed its reactions by catalyzing them.

Those who wished to believe that life began with nucleic acids, however, were faced with an inconvenient fact: nude DNA does not self-replicate. A complex assemblage of protein enzymes must already be in place. Subsequent work by biochemists Matthew Messelson and Franklin Stahl showed that DNA in the chromosomes within cells does replicate much as its structure suggests. Watson does specify a new Crick; Crick does specify a new Watson. But the cellular dance is mediated by a host of protein enzymes.

Those seeking the first living molecule would have to look elsewhere. And soon another polymer hove into the biologists' sight. RNA, or ribonucleic acid, is the first cousin of DNA and is central to the functioning of a cell. Like DNA, RNA is a polymer of four nucleotide bases: A, C, and G, as in DNA, but with uracil, U, substituting for thymine. RNA can exist as a single-stranded form or as a double helix. Like DNA, the two strands of the double helix of RNA are template complements. In the cell, the information to make a protein is copied from the DNA to a strand of so-called messenger RNA and is ferried to structures called ribosomes. At these sites, with the help of another kind of RNA molecule, transfer RNA, protein is fabricated.

The template complementarity of double-stranded RNA suggested to many scientists that RNA might be capable of replicating itself without the help of protein enzymes. Thus life might have begun as proliferating molecules of RNA—nude genes, as they are sometimes called. Perhaps sadly, the efforts to get RNA strands to copy themselves in a test tube have failed. But the idea is simple and beautiful: place in a beaker a high concentration of a specific single-stranded sequence—say, the decanucleotide CCCCCCCCCC. In addition, place a high concentration of free G nucleotides. Each G should line up with one of the C nucleotides in the decanucleotide, by virtue of Watson–Crick base pairing, such that a set of 10 G monomers are lined up adjacent to one another. It only remains that the 10 G monomer nucleotides become joined to one another by the proper bonds. Then what molecular biologists call a polyG decamer would have been formed. Then the two strands, polyG and polyC, need only melt apart, leaving the initial polyC decamer free to line up 10 more G monomers and create yet another polyG decamer. And finally, of course, to obtain a reproducing system of molecules, one hopes that the newly created polyG decamers, GGGGGGGGGG, might in turn be able to line up any free C monomers added to the beaker and cause the free C monomers to link together to form a polyC decamer, CCCCCCCCCC. Were all that to occur, and do so in the absence of any enzymes, then such a double-stranded RNA molecule would indeed be a nude replicating RNA molecule. Such a molecule would be a powerful candidate to be the first living molecule.

The idea is crisp and lovely. Almost without exception, however, the experiment does not work. The ways it fails are very instructive. First, each of the four nucleotides has its own chemical personality, which tends to make the experiment fail. Thus single-stranded polyG has a tendency to curl back on itself in a hairpin so that two G nucleotides bond to each other. The result is a tangled mess that is incapable of acting as a template for self-copying. Starting with a sequence of C and G monomers, richer in C than in G, a complementary string can be easily created. But that complement is necessarily richer in G than in C, and thus tends to curl back on itself, removing itself from the game. Watson makes Crick. Crick examines his navel and refuses to play.

Even if copying were not brought to a halt by guanine tangles, naked RNA molecules could suffer from what is called an error catastrophe: in copying one strand into another, misplaced bases—a G where a C should be—will corrupt the genetic message. In cells, these mistakes are kept to a minimum by proofreading enzymes that ensure faithful copies. Those rare mistakes that slip through the net are the mutations that drive evolution; most are detrimental, but occasionally one nudges the organism into a slightly fitter state. But without enzymes to avoid guanine tangles, copying errors, and other mistakes, an RNA message on its own could quickly become nonsense. And where, in a world of pure RNA, would the enzymes come from?

Some of those who believe that life began with RNA seek ways around this problem. Perhaps, they argue, a simpler self-replicating molecule came before RNA, one that was not plagued with guanine tangles and other problems. No clear experimental work currently backs up this approach. Should it work, we would also confront the question of how evolution converted such simpler polymers to RNA and DNA.

If enzymes are absolutely necessary to replication, then those who believe that RNA came first—and this is the mainstream view—must seek ways that nucleic acids themselves can act as catalysts. A mere decade ago, most biologists, chemists, and molecular biologists held to the view that the catalytic molecules of the cell were exclusively protein enzymes and that both DNA and RNA were essentially chemically inert storehouses of information. A faint whiff that RNA might be of more dynamic importance might have been scented in the fact that special RNA molecules, called transfer RNAs, play a critical role that can hardly be considered passive in translating the genetic code into proteins. Furthermore, the ribosome, the molecular machine in the cell that accomplishes the translation, is made up largely of RNA sequences plus some proteins. This machinery is almost identical throughout the living world and thus probably existed almost from the beginning of the time when matter became alive. But it was not until the mid-1980s that Thomas Cech and his colleagues made the stunning discovery that RNA molecules themselves can act as enzymes and catalyze reactions. Such RNA sequences are called ribozymes.

When the DNA message—the instructions for making a protein—is copied onto a strand of messenger RNA, a certain amount of information is ignored. And so cells have not only proofreading enzymes, but editing enzymes. The part of the sequence that contains the genetic instructions, the exons, must be separated from the nonsense sections, the introns. And so enzymes are used to snip out the introns from the RNA and splice together the exons. The sequence of now adjacent exons is processed in still other ways, is transported from the nucleus, finds a ribosome, and is translated into a protein. Cech discovered, undoubtedly with some astonishment, that in some cases no protein enzyme is needed for the editing. The RNA sequence itself acts as an enzyme, cutting out its own introns. The results rather flabbergasted the molecular biology community. It now turns out that a variety of such ribozymes exist and can catalyze a variety of reactions, acting on themselves or on other RNA sequences. For example, one is able to transfer a C nucleotide from the end of one sequence to the end of another: (CCCC) + (CCCC) yields (CCC) + (CCCCC).

RNA molecules, in the absence of protein enzymes, appear rather gauche at self-reproduction. But perhaps an RNA ribozyme might act as an enzyme and catalyze the reproduction of RNA molecules. And perhaps such a ribozyme might act on itself to reproduce itself. Either way, a self-reproducing molecule, or system of molecules, would lie to hand. Life would be under way.

It is important to be clear what is being requested of this ribozyme. Genetic information is carried in the sequence of nucleotide bases. Thus CCC is different from UAG. If one strand of an RNA molecule is UAGGCCUAAUUGA, then as its complementary strand is synthesized, the growing new sequence should be AUCCGGAUUAACU. As each new nucleotide is added, the proper choice must be made among the four possible nucleotides, and the proper bond must be formed. Protein enzymes are able to accomplish this fine discrimination and are called polymerases. RNA and DNA polymerases are utterly essential to the synthesis of RNA and DNA sequences within the cell. But it may not be easy to find an RNA sequence able to perform this polymerase function. Nevertheless, such a ribozyme polymerase is fully plausible. Such a molecule may have been present at the dawn of life.

And, again, maybe it was not. Indeed, a serious problem assails the ribozyme polymerase hypothesis. Grant that such a fine molecule arose. Could it sustain itself against mutational degradation? And could it evolve? The answer to both questions seems likely to be no. The problem is a form of an error catastrophe, first described by the chemist Leslie Orgel in the context of the genetic code. Picture a ribozyme that is able to function as a polymerase and copy any RNA molecule, including itself. Given a supply of nucleotides, this ribozyme would constitute a nude replicating gene. But any enzyme only hastens the correct reaction among the alternative possible side reactions that might also occur. Errors are inevitable. The self-reproducing ribozyme would necessarily produce mutant variants. But those mutant variant ribozymes themselves are likely to be less efficient than the normal, or wild-type, ribozyme, and hence are likely to make errors more frequently. These sloppier ribozymes will tend to reproduce themselves with even more mutants per copy than the wild-type ribozyme. Worse, the sloppy mutant ribozymes are able to catalyze the reproduction of the wild-type ribozyme, creating still more mutants. Over cycles, the system could produce a runaway spectrum of mutant variants. If so, the original ribozyme, with its ability to faithfully copy itself and others, could be lost in a flurry of sloppy catalysis leading to a system of RNA sequences that are catalytically inert. Life would have vanished in a runaway error catastrophe. I do not know of a detailed analysis of this specific problem, but I think that the potential for an error catastrophe for such a self-reproducing ribozyme deserves real analysis and encourages some caution about an otherwise very attractive hypothesis.

Of all the problems with the hypothesis that life started as nude replicating RNA molecules, the one I find most insurmountable is the one most rarely talked about: all living things seem to have a minimal complexity below which it is impossible to go. The simplest free-living cells are called pleuromona, a highly simplified kind of bacterium, replete with cell membrane, genes, RNA, protein-synthesizing machinery, proteins—the full complement of standard gear. The number of genes in pleuromona is variously estimated at a few hundred to about a thousand, compared with the estimated 3,000 in *Escherichia coli,* a bacterium in our intestines. Pleuromona is the simplest thing that we know to be alive. Your curiosity should be aroused. Viruses, which are vastly simpler than the pleuromona, are not free living. They are parasites that invade cells, co-opt the cell's metabolic machinery to accomplish their own self-reproduction, escape the host cell, and invade another. All free-living cells have at least the minimum molecular diversity of pleuromona. Your antenna should quiver a bit here. Why is there this minimal complexity? Why can't a system simpler than pleuromona be alive?

The best answer that the advocates of an RNA world can offer is an evolutionary just-so story, in honor of Rudyard Kipling and his fanciful tales about how different animals came to be. In medical school, I learned that a bone filled with small holes, fenestrated, and called the cribiform plate, forms the junction of the nose and forehead. The evolutionary rationale of this bone was explained to us: light and strong, well adapted to its function. Now, had the cribiform plate been a solid chunk of knobby bone creating a horny protrusion, a kind of awning to my proboscis, I have no doubt that our professor would again have found a use for this solid massive knob, highly adapted for banging one's head against the wall. Evolution is filled with these just-so stories, plausible scenarios for which no evidence can be found, stories we love to tell but on which we should place no intellectual reliance.

What would Kipling (or, for that matter, most evolutionary biologists) say about why simple RNA replicators gave rise to a world where life seems to occur only above a threshold of complexity? Because, comes the answer, these first living molecules, driven by mutation and the survival of the fittest, gathered about themselves the vestments of metabolism, membrane, and so forth. Eventually, they evolved into the cells we have today. Fully clothed, the current minimal cell happens to have the observed minimal complexity. But there is nothing deep about this explanation. We have instead another just-so story, plausible but not convincing, and, as with all just-so stories, the implication is that things could easily be another way. If the chain of accidents leading to our emergence had taken another route, we might indeed have horny protrusions on our foreheads. If RNA molecules had ascended according to a different route, the threshold of complexity might be different: things simpler than pleuromona might be able to sustain themselves. Or, alternatively, the simplest possible life form might be the mollusk.

In short, the nude RNA or the nude ribozyme polymerase offers no deep account of the observed minimal complexity of all free-living cells. I hold it as a virtue of the origins theory I shall describe in Chapter 3 that it makes it clear why matter must reach a certain level of complexity in order to spring into life. This threshold is not an accident of random variation and selection; I hold that it is inherent to the very nature of life.

The Crystallization of Life

We are not supposed to be here. Life cannot have occurred. Before you get up to leave your chair, your very existence standing as blunt refutation to the argument you are about to hear, simple intellectual politeness invites you to reconsider and linger awhile. The argument I now present has been held seriously by very able scientists. Its failure, I believe, lies in its inability to understand the profound power of self-organization in complex systems. I shall be at pains to show you soon that such self-organization may have made the emergence of life well-nigh inevitable.

We begin on an optimistic note with an argument for the origin of life on earth from the Nobel laureate George Wald in an article in *Scientific American* in 1954. Wald wonders how it could be that a collection of molecules came together in just the right way to form a living cell. One has only to contemplate the magnitude of this task to concede that the spontaneous generation of a living organism is impossible. Yet we are here. Wald goes on to argue that, with very many trials, the unthinkably improbable becomes virtually assured. Time is in fact the hero of the plot. The time with which we have to deal is of the order of 2 billion years. (Wald wrote in 1954; we now say 4 billion years.) Given so much time, the impossible becomes possible, the possible probable, and the probable virtually certain. One has only to wait; time itself performs the miracles.

But critics arose, critics of high renown, to argue that even 2 or 4 billion years was not enough time for life to arise by pure happenstance, not by vast orders of magnitude. In his book *Origins,* Robert Shapiro calculates that in the history of the earth, there could conceivably have been 2.5×10^{51} attempts to create life by chance. That is one hell of a lot of trials. But is it enough? We need to know the probability of success per trial.

Shapiro continues with an effort to calculate the odds of attaining, by chance, something like *E. coli.* He begins with an argument by two astronomers, Sir Fred Hoyle and N. C. Wickramasinghe. Rather than estimate the chances for obtaining an entire bacterium, these authors try to calculate the chances for obtaining a functioning enzyme. They begin with the set of 20 amino acids that are used to construct enzymes. If the amino acids were selected at random and arranged in random order, what would be the chances of obtaining an actual bacterial enzyme with 200 amino acids? The answer is obtained by multiplying the probability for each correct amino acid in the sequence, 1 in 20, together 200 times, yielding 1 in 20^{200}, a vastly low probability. But since more than one amino acid sequence might be able to function to catalyze a given reaction, the authors concede a probability of 1 in 10^{20}. But now the coup de grâce: to duplicate a bacterium, it would not suffice to create a single enzyme. Instead, it would be necessary to assemble about 2,000 functioning enzymes. The odds against this would be 1 in $10^{20 \times 2,000}$, or 1 in $10^{40,000}$. These exponential notations are easy to state, but difficult to take to heart. The total number of hydrogen atoms in the universe is something like 10^{60}. So $10^{40,000}$ is vast beyond vast, unimaginably hyperastronomical. And 1 in $10^{40,000}$ is unthinkably improbable. If the total number of trials for life to get going is only 10^{51}, and the chances are 1 in $10^{40,000}$, then life just could not have occurred. We the lucky. We the very, very lucky. We the impossible. Hoyle and Wickramasinghe gave up on spontaneous generation, since the likelihood of the event was comparable to the chances that a tornado sweeping through a junkyard might assemble a Boeing 747 from the materials therein.

Since you are reading this book, and I am writing it, something must be wrong with the argument. The problem, I believe, is that Hoyle, Wickramasinghe, and many others have failed to appreciate the power of self-organization. It is not necessary that a specific set of 2,000 enzymes be assembled, one by one, to carry out a specific set of reactions. As we shall see in Chapter 3, there are compelling reasons to believe that whenever a collection of chemicals contains enough different kinds of molecules, a metabolism will crystallize from the broth. If this argument is correct, metabolic networks need not be built one component at a time; they can spring full-grown from a primordial soup. Order for free, I call it. If I am right, the motto of life is not We the improbable, but We the expected.

Chapter 3

We the Expected

W hat raw day first saw life, raw itself, pregnant with the future? Four billion years, thereabouts, from the first circle of metabolic witchery to me and to thee. Raw chance? Raw improbability that ought never have occurred in billions of times the history of this universe? Raw meaninglessness that we are so very unexplained?

Is life really the unthinkable accident that follows from the calculations of Fred Hoyle and N. C. Wickramasinghe? Is time the hero of the plot, as George Wald argued? Yet we now believe there were but 300 million years or so from the cooling of the crust to clear evidence of cellular life, not the 2 billion years that Wald appealed to. Time was not there in sufficient vastness for Wald's story, and surely not for Hoyle and Wickramasinghe's tale. If we the living are wildly improbable, then we are unaccountable mysteries in the span of space and time. But if this view is wrong, if there is some reason to believe that life is probable, then we are not mysteries in the exploding cosmos, we are natural parts of it.

Most of my colleagues believe that life emerged simple and became complex. They picture nude RNA molecules replicating and replicating and eventually stumbling on and assembling all the complicated chemical machinery we find in a living cell. Most of my colleagues also believe that life is utterly dependent on the molecular logic of template replication, the A–T, G–C Watson–Crick pairing that I wrote about in Chapter 2. I hold a renegade view: life is not shackled to the magic of template replication, but based on a deeper logic. I hope to persuade you that life is a natural property of complex chemical systems, that when the number of different kinds of molecules in a chemical soup passes a certain threshold, a self-sustaining network of reactions—an autocatalytic metabolism—will suddenly appear. Life emerged, I suggest, not simple, but complex and whole, and has remained complex and whole ever since—not because of a mysterious élan vital, but thanks to the simple, profound transformation of dead molecules into an organization by which each molecule's formation is catalyzed by some other molecule in the organization. The secret of life, the wellspring of reproduction, is not to be found in the beauty of Watson–Crick pairing, but in the achievement of collective catalytic closure. The roots are deeper than the double helix and are based in chemistry itself. So, in another sense, life—complex, whole, emergent—is simple after all, a natural outgrowth of the world in which we live.

The claim that life emerges as a natural phase transition in complex chemical systems is so radical a departure from past theories that I owe you caveats. Do we know that such a view is at least theoretically coherent? Do we know it to be physically and chemically possible? Is there evidence for such a view? Is evidence attainable? Do we know that life began as I shall suggest it did? The most that can be said at this stage is that good, careful theoretical work strongly supports the possibility I shall present. That work appears to be consistent with what we know about complex chemical systems. Scant experimental evidence supports this view as yet, but stunning developments in molecular biology now make it possible to imagine actually creating these self-reproducing molecular systems—synthesized life. I believe that this will be accomplished within a decade or two.

The Networks of Life

As noted in Chapter 2, most researchers are focusing their attention on the capacity of RNA, or RNA-like polymers, to self-reproduce by template replication. The attention is understandable. No one looking at the beautiful double helix of DNA or RNA and regarding the Watson–Crick pairing rules can avoid being struck by the beauty of nature's apparent choice. The fact that Leslie Orgel and his colleagues have not yet succeeded in getting such polymers to replicate without an enzyme does not mean that the efforts will always fail. Orgel has been at it for perhaps 25 years; nature took something like 100 million years. Orgel is very smart, but 100 million years is long enough, measured in three-year National Institutes of Health grants, to try lots of possibilities. Let us try a different tack. Suppose that the laws of chemistry were slightly different, that nitrogen had four rather than five valence electrons, say, allowing four rather than five bonding partners. Ignore the wrench this would throw into quantum mechanics—one can sometimes get away

with being wretched to quantum mechanics when making a philosophi-
cal point. If the laws of chemistry were slightly different so that the
beautiful double-helix structure of DNA and RNA were no longer possible, would life based on chemistry be impossible? I do not want to
think that we were quite so lucky. I hope we can find a basis for life that
lies deeper than template self-complementarity.

The secret, I believe, lies in what chemists call catalysis. Many chemical reactions proceed only with great difficulty. Given a long expanse of
time, a few molecules of A might combine with molecules of B to make
C. But in the presence of a catalyst, another molecule we'll call D, the
reaction catches fire and proceeds very much faster. The usual
metaphor is the lock and key: A and B fit into slots on D, in just such a
way that they are far more likely to combine to form C. As we shall see,
this is a vast oversimplification, but for now it will suffice to get the
point across. While D is the catalyst that joins A and B to make C, the
molecules A, B, and C might themselves act as catalysts for other reactions.

At its heart, a living organism is a system of chemicals that has the capacity to catalyze is own reproduction. Catalysts such as enzymes speed
up chemical reactions that might otherwise occur, but only extremely
slowly. What I call a collectively autocatalytic system is one in which the
molecules speed up the very reactions by which they themselves are
formed: A makes B; B makes C; C makes A again. Now imagine a whole
network of these self-propelling loops (Figure 3.1). Given a supply of

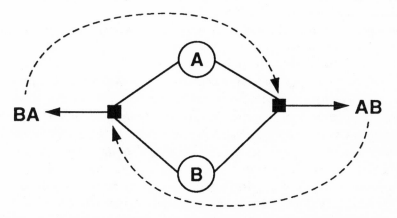

Figure 3.1 *A simple autocatalytic set. Two dimer molecules, AB and BA, are
formed from two simple monomers, A and B. Since AB and BA catalyze the very
reactions that join As and Bs to make the dimers, the network is autocatalytic:
given a supply of "food" molecules (As and Bs), it will sustain itself.*

food molecules, the network will be able to constantly re-create itself. Like the metabolic networks that inhabit every living cell, it will be alive. What I aim to show is that if a sufficiently diverse mix of molecules accumulates somewhere, the chances that an autocatalytic system—a self-maintaining and self-reproducing metabolism—will spring forth becomes a near certainty. If so, then the emergence of life may have been much easier than we have supposed.

What I aim to show is simple, but radical. I hold that life, at its root, does not depend on the magic of Watson–Crick base pairing or any other specific template-replicating machinery. Life, at its root, lies in the property of *catalytic closure* among a collection of molecular species. Alone, each molecular species is dead. Jointly, once catalytic closure among them is achieved, the collective system of molecules is alive.

Each cell in your body, every free-living cell, is collectively autocatalytic. No DNA molecules replicate nude in free-living organisms. DNA replicates only as part of a complex, collectively autocatalytic network of reactions and enzymes in cells. No RNA molecules replicate themselves. The cell is a whole, mysterious in its origins perhaps, but not mystical. Except for "food molecules," every molecular species of which a cell is constructed is created by catalysis of reactions, and the catalysis is itself carried out by catalysts created by the cell. To understand the origin of life, I claim, we must understand the conditions that enabled the first emergence of such autocatalytic molecular systems.

Catalysis alone, however, is not sufficient for life. All living systems "eat": they take in matter and energy in order to reproduce themselves. This means that they are what is referred to in Chapter 1 as open thermodynamic systems.

In contrast, closed thermodynamic systems take in no matter or energy from their environments. A great deal is understood about the behavior of closed thermodynamic systems. The theorists of thermodynamics and statistical mechanics have studied such systems for over 100 years. In contrast, remarkably little is understood about the possible behaviors of open thermodynamic systems. Not so surprising, this ignorance. The vast flowering of all life-forms over the past 3.45 billion years is merely a hint of the possible behaviors of open thermodynamic systems. So too is cosmogenesis itself, for the evolving universe since the Big Bang has yielded the formation of galactic and supragalactic structures on enormous scales. Those stellar structures and the nuclear processes within stars, which have generated the atoms and molecules from which life itself arose, are open systems, driven by nonequilibrium processes. We have only begun to understand the awesome creative powers of nonequilibrium processes in the unfolding universe. We are all—complex atoms, Jupiter, spiral galaxies, warthog, and frog—the logical progeny of that creative power.

Since I hope to persuade you that life is the natural accomplishment of catalysts in sufficiently complex nonequilibrium chemical systems, I had best take a moment to sketch what catalysts accomplish and how equilibrium and nonequilibrium chemical systems behave. Chemical reactions occur spontaneously, some rapidly, some slowly. Typically, chemical reactions are more or less reversible: A transforms to B, but B transforms to A. Since such reactions are reversible, it is easy to think about what would occur in a beaker that began with an initial concentration of A molecules and no B molecules and that was closed to the addition of matter or energy. The A molecules would begin to convert to B molecules, but as that occurred, the new B molecules would begin to convert back to A molecules. Starting with only A molecules, the B concentration would build up to the point at which the rate of conversion of A to B was exactly equal to the rate of conversion of B to A. This balance is called chemical equilibrium. At chemical equilibrium, the net concentrations of A and B do not change over time, but any given A molecule may convert to B and back again thousands of times per minute. Of course, the equilibrium is statistical. Minor fluctuations in A and B concentrations occur all the time.

Chemical equilibrium is not limited to a pair of molecules, A and B, but will occur in any closed thermodynamic system. If the system has hundreds of different types of molecules, it will ultimately settle down to an equilibrium in which the forward and reverse reactions between any pair of molecules balance out.

Catalysts, of which protein enzymes and ribozymes are examples, can speed up both the forward and the reverse reaction by the same amount. The equilibrium between A and B is not altered; enzymes simply hasten the rate at which this state of balance is reached. Suppose, at equilibrium, the ratio of A and B concentrations is 1, so the concentrations of the two are equal. If the chemical system starts out displaced from equilibrium—say, with a high concentration of B and almost no A—then the enzyme will vastly shorten the time it takes to reach the equilibrium ratio where the two concentrations are equal. In effect, then, the enzyme increases the rate of production of A.

How does catalysis happen? There is an intermediate state between A and B, called the transition state, in which one or more bonds among the atoms of the molecule are severely strained and distorted. The transition-state molecule is therefore rather unhappy. The measure of this unhappiness is given by the energy of the molecule. Low energy corresponds to unstrained molecules. High energy corresponds to strained molecules. Think of a spring. At its rest length, it is happy. If stretched beyond its rest length, it has stored energy—it is unhappy—and can release that energy by snapping back to its rest length, whereupon it has low energy again.

Not surprisingly, the transition state passing from A to B is exactly the same as the transition state passing from B back to A. Enzymes are thought to work by binding to the transition state and stabilizing it. This makes it easier for both A and B molecules to jump to the transition state, increasing the rate of conversion of A to B, and of B to A. Thus an enzyme increases the rate at which the equilibrium ratio of A and B concentrations is approached.

We should be thankful that our cells are not at chemical equilibrium; for a living system, equilibrium corresponds to death. Living systems are, instead, open thermodynamic systems persistently displaced from chemical equilibrium. We eat and excrete, as did our remote ancestors. Energy and matter flow through us, building up the complex molecules that are the tokens in the game of life.

Open nonequilibrium systems obey very different rules from those of closed systems. Consider a simple case: we have a beaker into which we add A molecules continuously from some outside source at a constant rate, and we take any B molecules out of the beaker at a rate proportional to the concentration of B. A will convert to B and B will convert to A as before, but the two molecules can never reach the equilibrium balance they attained before because of the constant addition of A and the removal of B. Common sense says that the system will settle down to a steady state at which the ratio of A molecules to B molecules is higher than it was when the system was closed. In short, the ratio of A to B will be tipped from the thermodynamic equilibrium ratio. In general, this commonsense view is correct. In simple cases, such systems, open to the flux of matter and energy, settle down to a steady state different from that found in closed thermodynamic systems.

Now consider a vastly more complex open system, the living cell. The cells of your body coordinate the behaviors of about 100,000 different kinds of molecules as matter and energy cross their boundaries. Even bacteria coordinate the activities of thousands of different kinds of molecules. To think that understanding the behavior of very simple open thermodynamic chemical systems takes us far toward understanding the cell is hubris. No one understands how the complex cellular networks of chemical reactions and their catalysts behave, or what laws might govern their behavior. Indeed, this is a mystery we will begin to discuss in the next chapter. Yet simple open thermodynamic systems are at least a start and are already fascinating on their own. Even simple nonequilibrium chemical systems can form remarkably complex patterns of chemical concentrations varying in time and space in striking ways. As noted in Chapter 1, Ilya Prigogine called these systems dissipative because they persistently dissipate matter and energy in order to maintain their structures.

Unlike the simple steady-state system in the thermodynamically open beaker, the concentrations of the chemical species in a more complex dissipative system may not fall to a steady state, unchanging in time. Instead, the concentrations can start to oscillate up and down in repeated cycles, called limit cycles, which are sustained for long periods of time. Such systems can also generate remarkable spatial patterns. For example, the famous Belosov–Zhabotinski reaction, made of some simple organic molecules, sets up two kinds of spatial patterns. In the first pattern, spreading concentric circular waves of blue propagate outward over an orange background from a central oscillating source. The blue and orange colors arise because of indicator molecules that track how acidic or basic the reaction mixture is at any point in space. In the second pattern, spiral pinwheels of blue on orange cartwheel about a center (Figure 3.2). Such patterns have been studied by a number of researchers. A fine book by my friend Arthur Winfree, *When Time Breaks Down: The Three-Dimensional Dynamics of Electrochemical Waves and Cardiac Arrhythmias,* summarizes much of the work. Among the most immediate human implications is this: the heart is an open system, and it can beat according to patterns analogous to the Belosov–Zhabotinski

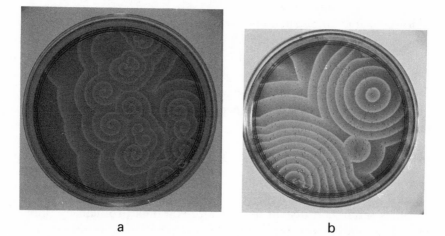

a   b

Figure 3.2 *Self-organization at work. The famous Belosov–Zhabotinski reaction showing the spontaneous emergence of order in a simple chemical system. (a) Concentric circular waves propagate outward. (b) Radially expanding pinwheels cartwheel about a center.*

reaction. Sudden death caused by cardiac arrhythmias may correspond to a switch from the analogue of the concentric-circles pattern (a steady beating) to the spiral-pinwheels pattern in your myocardium. The blue propagating wave can be thought of as corresponding to the chemical conditions in muscle cells that lead them to contract. Thus the concentric spreading pattern of the evenly spaced blue circles corresponds to ordered contraction waves. But in the spiral pattern, the blue pinwheels are very close together near the center of the spiral and are spaced farther apart the farther out on the spiral they go. This pattern corresponds to chaotic twitching of the heart muscle in the vicinity of the spiral center. Winfree has shown that simple perturbations, such as shaking the petri plate that holds the chemical reactants of the Belosov–Zhabotinski reaction, can switch the system from the concentric to the spiral pattern. Thus Winfree has suggested that simple perturbations can switch a normal heart to the spiral chaotic pattern and lead to sudden death.

The relatively simple behaviors of nonequilibrium chemical systems are well studied and may have a variety of biological implications. For example, such systems can form a standing pattern of stripes of high chemical concentrations spaced between stripes of low chemical concentrations. Many of us think that the natural patterns such systems form have a great deal to tell us about the spatial patterning that occurs in the development of plants and animals. The blue and orange stripes in the Belosov–Zhabotinski reaction may foretell the stripes of the zebra, the banding patterns on shells, and other aspects of morphology in simple and complex organisms.

However intriguing such chemical patterns may be, they are not yet living systems. The cell is not only an open chemical system, but a collectively autocatalytic system. Not only do chemical patterns arise in cells, but cells sustain themselves as reproducing entities that are capable of Darwinian evolution. By what laws, what deep principles, might autocatalytic systems have emerged on the primal earth? We seek, in short, our creation myth.

A Chemical Creation Myth

Scientists often gain insight into a more complex problem by thinking through a simpler toy problem. The toy problem I want to tell you about concerns "random graphs." A random graph is a set of dots, or nodes, connected at random by a set of lines, or edges. Figure 3.3 shows an example. To make the toy problem concrete, we can call the dots

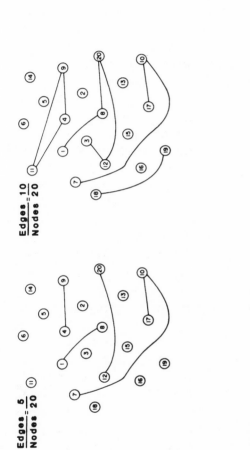

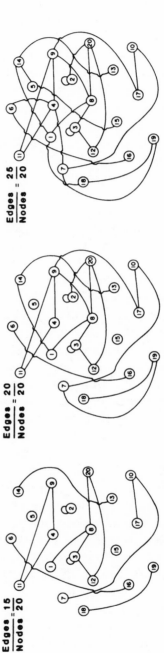

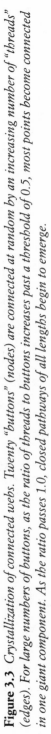

Figure 3.3 *Crystallization of connected webs. Twenty "buttons" (nodes) are connected at random by an increasing number of "threads" (edges). For large numbers of buttons, as the ratio of threads to buttons increases past a threshold of 0.5, most points become connected in one giant component. As the ratio passes 1.0, closed pathways of all lengths begin to emerge.*

"buttons" and the lines "threads." Imagine 10,000 buttons scattered on a hardwood floor. Randomly choose two buttons and connect them with a thread. Now put this pair down and randomly choose two more buttons, pick them up, and connect them with a thread. As you continue to do this, at first you will almost certainly pick up buttons that you have not picked up before. After a while, however, you are more likely to pick at random a pair of buttons and find that you have already chosen one of the pair. So when you tie a thread between the two newly chosen buttons, you will find three buttons tied together. In short, as you continue to choose random pairs of buttons to connect with a thread, after a while the buttons start becoming interconnected into larger clusters. This is shown in Figure 3.3a, which is limited to 20 rather than 10,000 buttons. Every now and then, lift up a button and see how many other buttons you pick up. The connected cluster is called a component in our random graph. As Figure 3.3a shows, some buttons may not be connected to any other buttons. Other buttons might be connected in pairs or triples or larger numbers.

The important features of random graphs show very regular statistical behavior as one tunes the ratio of threads to buttons. In particular, a *phase transition* occurs when the ratio of threads to buttons passes 0.5. At that point, a "giant cluster" suddenly forms. Figure 3.3 shows this process, using only 20 buttons. When there are very few threads compared with the number of buttons, most buttons will be unconnected (Figure 3.3a), but as the ratio of threads to buttons increases, small connected clusters begin to form. As the ratio of threads to buttons continues to increase, the size of these clusters of buttons tends to grow. Obviously, as clusters get larger, they begin to become cross-connected. Now the magic! As the ratio of threads to buttons passes the 0.5 mark, all of a sudden most of the clusters have become cross-connected into one giant structure. In the small system with 20 buttons in Figure 3.3, you can see this giant cluster forming when the ratio of threads to buttons is half, 10 threads to 20 buttons. If we used 10,000 buttons, the giant component would arise when there were about 5,000 threads. When the giant component forms, most of the nodes are directly or indirectly connected. If you pick up one button, the chances are high that you will pull up something like 8,000 of the 10,000 buttons. As the ratio of threads to buttons continues to increase past the halfway mark, more and more of the remaining isolated buttons and small clusters become cross-connected into the giant component. So the giant component grows larger, but its rate of growth slows as the number of remaining isolated buttons and isolated small components decreases.

The rather sudden change in the size of the largest connected cluster of buttons, as the ratio of threads to buttons passes 0.5, is a toy version of the phase transition that I believe led to the origin of life. In Figure 3.4, I show qualitatively the size of the largest cluster among 400 nodes as the ratio of edges to nodes increases. Note that the curve is S-shaped, or sigmoidal. The size of the largest cluster of nodes increases slowly at first, then rapidly, then slows again as the ratio of edge to nodes increases. The rapid increase is the signature of something like a phase transition (Figure 3.4). In the example in Figure 3.4 using 400 buttons, the sigmoidal curve rises steeply when the ratio of edges to nodes passes 0.5. The steepness of the curve at the critical 0.5 ratio depends on the number of nodes in the system. When the number of nodes is small, the steepest part of the curve is "shallow," but as the number of nodes in the toy system increases—from, say, 400 to 100 million—the steep part of the sigmoidal curve becomes more vertical. Were there an infinite number of buttons, then as the ratio of threads to buttons passed 0.5 the size of the largest component would jump discontinuously from tiny to enormous. This is a phase transition, rather like separate water molecules freezing into a block of ice.

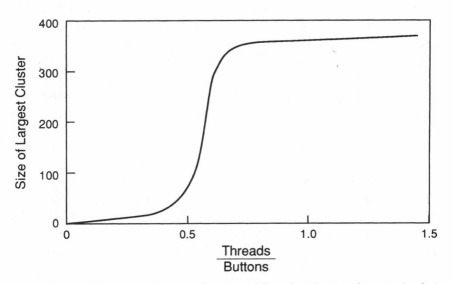

Figure 3.4 *A phase transition. As the ratio of threads (edges) to buttons (nodes) in a random graph passes 0.5, the size of the connected cluster slowly increases until it reaches a "phase transition" and a giant component crystallizes. (For this experiment, the number of threads ranges from 0 to 600, while the number of buttons is fixed at 400.)*

The intuition I want you to take away from this toy problem is simple: as the ratio of threads to buttons increases, suddenly so many buttons are connected that a vast web of buttons forms in the system. This giant component is not mysterious; its emergence is the natural, expected property of a random graph. The analogue in the origin-of-life theory will be that when a large enough number of reactions are catalyzed in a chemical reaction system, a vast web of catalyzed reactions will suddenly crystallize. Such a web, it turns out, is almost certainly autocatalytic—almost certainly self-sustaining, alive.

Reaction Networks

It is convenient to draw a metabolic reaction graph with circles representing chemicals and square representing reactions. To be concrete, we will consider four simple kinds of reactions. In the simplest, one substrate, A, converts to one product, B. Since reactions are reversible, B also converts back to A. This is a one-substrate, one-product reaction. Draw a black line leaving A and entering a small square lying between A and B, and draw a line leaving the square and ending on B (Figure 3.5). This line and the square represent the reaction between A and B. Now consider two molecules, say A and B, that are combined, or "ligated," to form a larger molecule, C. In the reverse reaction, C is "cleaved" to form A and B. We can represent these reactions with two lines leaving A and B and entering a square representing this reaction, plus a line leaving the square and entering C. Finally, we should consider reactions with two substrates and two products. Typically, this kind of reaction occurs by breaking off a small cluster of atoms from one substrate and bonding the cluster to one or more atoms on the second substrate. We can represent two-substrate, two-product reactions with pairs of lines leaving the two substrates and entering a square representing that reaction, and two more lines leaving the square and connecting to the two products. Now consider all the kinds of molecules and reactions possible in a chemical reaction system. The collection of all such lines and squares between all the chemical circles constitutes the reaction graph (Figure 3.5).

Since we want to understand the emergence of collectively autocatalytic molecular systems, the next step is to distinguish between spontaneous reactions, which are assumed to occur very slowly, and catalyzed reactions, which are assumed to occur rapidly. We want to find the conditions under which the same molecules will be catalysts for and products of the reactions creating the autocatalytic set. This depends on the

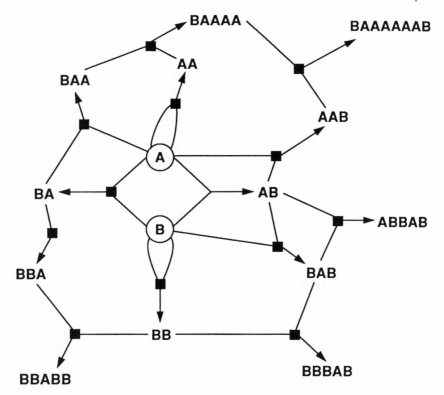

Figure 3.5 *From buttons and threads to chemicals. In this hypothetical network of chemical reactions, called a reaction graph, smaller molecules (A and B) are combined to form larger molecules (AA, AB, etc.), which are combined to form still larger molecules (BAB, BBA, BABB, etc.). Simultaneously, these longer molecules are broken down into simple substrates again. For each reaction, a line leads from the two substrates to a square denoting the reaction; an arrow leads from the reaction square to the product. (Since reactions are reversible, the use of arrows is meant to distinguish substrates from products in only one direction of the chemical flow.) Since the products of some reactions are substrates of further reactions, the result is a web of interlinked reactions.*

possibility that each molecule in the system can play a double role: it can serve as an ingredient or a product of a reaction, but it can also serve as a catalyst for another reaction. This dual role, as ingredient or catalyst, is perfectly possible, even familiar. Proteins and RNA molecules are known to play such a dual role. An enzyme called trypsin cleaves proteins you eat into smaller fragments. In fact, trypsin will cleave itself into fragments as well. And, as noted in Chapter 2, ribozymes are RNA molecules that can act as enzymes on RNA molecules. It is perfectly familiar that all kinds of organic molecules can be substrates and products of reactions, but simultaneously act catalytically to hasten other reactions. No mystery stands in the way of a dual role for chemicals.

To proceed further, we need to know which molecules catalyze which reactions. If we knew this, we could tell whether any set of molecules might be collectively autocatalytic. Unfortunately, this knowledge is not, in general, yet available, but we can sensibly proceed by making plausible assumptions. I will consider two such simple theories, each of which allows us, in the model worlds we will consider, to assign, somewhat arbitrarily, catalysts to reactions. You should be skeptical about this maneuver. Surely, it might be thought, one must actually know which molecules catalyze which reactions to be certain that a set of molecules harbors an autocatalytic set. Such skepticism is well placed and allows me to introduce a mode of reasoning on which I am depending. One might easily object that if in the real world of chemical reactions the laws of chemistry dictated a somewhat different distribution of which molecules catalyzed which reactions, then the conclusions would not hold. My response is this: if we can show that for many alternative "hypothetical" chemistries, in which different molecules catalyze different reactions, autocatalytic sets emerge, then the particular details of the chemistry may not matter. We will be showing that the spontaneous emergence of self-sustaining webs is so natural and robust that it is even deeper than the specific chemistry that happens to exist on earth; it is rooted in mathematics itself.

Picture, as noted earlier, a reaction between a pair of molecules, A and B, as black lines or edges connecting A and B to the reaction square between them. Now picture some other molecule, C, that is able to catalyze the reaction between A and B. Represent this by drawing a blue arrow with its tail in C and its head on the reaction square between A and B (Figure 3.6). Represent the fact that the reaction between A and B is catalyzed by changing the black line between A and B to a red line. Consider each molecule in the system, and ask which reaction or reactions, if any, it can catalyze. For any such catalyst, draw a blue arrow to the corresponding reaction square, and color the corresponding reaction edges red. When you have finished this task, the red edges and the chemical nodes they connect represent all the catalyzed reactions, and collectively make up the *catalyzed reaction subgraph* of the whole reaction graph. The blue arrows and the chemical nodes from which they leave represent the molecules that carry out the catalysis (Figure 3.6).

Now consider what is required for the system to contain an autocatalytic subset: first, a set of molecules must be connected by red catalyzed reactions; second, the molecules in this set must each have its

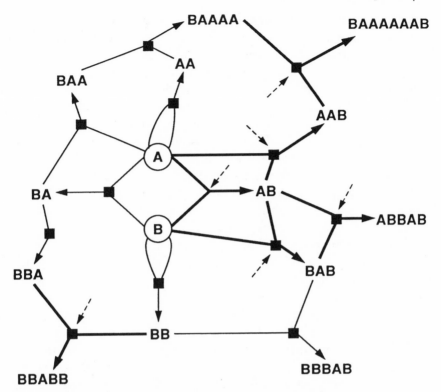

Figure 3.6 *Molecules catalyzing reactions. In Figure 3.5, all the reactions were assumed to be spontaneous. What happens when we add catalysts to speed some of the reactions? Here the reaction squares indicated by dashed-line arrows are catalyzed, and the heavy, darker lines connect substrates and products whose reactions are catalyzed. The result is a pattern of heavy lines indicating a catalyzed subgraph of the reaction graph.*

formation catalyzed by a blue arrow from some molecule in the same set or be added from outside. Call the latter molecules food molecules. If these conditions are met, we have a network of molecules that can catalyze its own formation, creating all the catalysts it needs.

The Central Idea

How likely is it that such a self-sustaining web of reactions would arise naturally? Is the emergence of collective autocatalysis easy or virtually impossible? Do we have to pick our chemicals carefully, or would just about any mixture do? The answer is heartening. The emergence of autocatalytic sets is almost inevitable.

Here, in a nutshell we will unpack later, is what happens: as the diversity of molecules in our system increases, the ratio of reactions to chemicals, or edges to nodes, becomes ever higher. In other words, the reaction graph has ever more lines connecting the chemical dots. The molecules in the system are themselves candidates to be able to catalyze the reactions by which the molecules themselves are formed. As the ratio of reactions to chemicals increases, the number of reactions that are catalyzed by the molecules in the system increases. When the number of catalyzed reactions is about equal to the number of chemical dots, a giant catalyzed reaction web forms, and a collectively autocatalytic system snaps into existence. A living metabolism crystallizes. Life emerges as a phase transition.

Now we will unpack our nutshell.

The first step is to show that as the diversity and complexity of the molecules in our system increase, the ratio of reactions to chemical dots in the reaction graph increases as well. It is easy to see why this is true. Consider a polymer consisting of four "monomers," which we can think of as atoms ABBB. Clearly, the polymer can be formed by gluing A to BBB, by gluing AB to BB, or by gluing ABB to B. So it can be formed in three ways, by three different reactions. If we increase the length of the polymer by one atom, the number of reactions per molecule will rise. ABBBA can be formed from A and BBBA, AB and BBA, ABB and BA, or ABBB and A. Since a polymer of length L has $L - 1$ internal bonds in general, a polymer of length L can be formed from smaller polymers in $L - 1$ ways. But these numbers account for only what chemists call ligation reactions, building up molecules from smaller pieces. Molecules can also be formed through cleavage. ABBB can be formed by lopping the A from the right-hand side of ABBBA. So it is rather obvious that there are more reactions by which molecules can be formed than there are molecules themselves. This means that in the reaction graph there are more lines than dots.

What happens to the ratio of reactions to molecules in the reaction graph as the diversity and complexity of those molecules increase? After some simple algebra, it is easy to show for simple linear polymers that as the length of the molecules increases, the number of kinds of molecules increases exponentially, but the number of reactions by which they convert from one to another rises even faster. This increasing ratio means that as more complex and diverse sets of molecules are considered, the reaction graph among them becomes ever denser with paths by which they can change from one into another. The ratio of reaction "lines" to dots becomes denser, a black forest of possibilities. The chemical system becomes ever more fecund with reactions by which molecules transform into other molecules.

At this point we have a flask of slow, spontaneous reactions. For the system to catch fire and generate self-sustaining autocatalytic networks, some of the molecules must act as catalysts, speeding up the reactions. The system is fecund, but not yet pregnant with life, and will not become so until we have a way to determine which molecules catalyze which reactions. Thus it is time to build some simple models. The simplest, which will do very well for a variety of purposes, is to assume that each polymer has a fixed chance, say one in a million, of being able to function as an enzyme to catalyze any given reaction. In using this simple model, we will "decide" which reactions, if any, each polymer can catalyze by flipping a biased coin that comes up heads once in a million times. Using this rule, any polymer will be randomly assigned, once and for all, the reactions it can catalyze. Using this "random catalyst" rule, we can "color" the catalyzed reactions red, draw our blue arrows from the catalysts to the reactions each catalyzes, and then ask whether our model chemical system contains a collectively autocatalytic set: a network of molecules connected by red lines and also containing the very molecules that catalyze, via the blue arrows, the reactions by which the molecules themselves are formed.

A somewhat more chemically plausible model supposes that our polymers are RNA sequences and introduces template matching. In this simplified version, Bs fit with As in a kind of Watson–Crick pairing. Thus the hexamer BBBBBB might be able to act like a ribozyme and bind two substrates, BABAAA and AAABBABA, by their two corresponding AAA trimer sites, and catalyze the ligation of the two substrates to form BABAAAAABBABA. To make things even more chemically realistic, we might also demand that even if a candidate ribozyme has a site that matches the left and right ends of its substrates, it still has only one chance in a million to have other chemical properties that allow it to catalyze the reaction. This captures the idea that other chemical features beyond template matching may be required to achieve ribozyme catalysis. Let us call this the match catalyst rule.

Here is the crucial result: no matter which of these "catalyst" rules we use, when the set of model molecules reaches a critical diversity, a giant "red" component of catalyzed reactions crystallizes, and so collectively autocatalytic sets emerge. Now it is easy to see why this emergence is virtually inevitable. Suppose we use the random catalyst rule and assume that any polymer has a one-in-a-million chance to act as an enzyme for any given reaction. As the diversity of molecules in the model system increases, the ratio of reactions to molecules increases. When the diversity of molecules is high enough, the ratio of reactions to polymers reaches a million to one. At that diversity, on average each polymer will catalyze one reaction. A million to one multiplied by one in a million equals one. When the ratio of catalyzed reactions to chemicals is 1.0, then with extremely high probability a "red" giant component, a web of catalyzed reactions, will form—a collectively autocatalytic set of molecules.

In this view of the origin of life, a critical diversity of molecules must be reached for the system to catch fire, for catalytic closure to be attained. A simple system with 10 polymers in it and a chance of catalysis of one in a million is just a set of dead molecules. Almost certainly, none of the 10 molecules catalyzes any of the possible reactions among the 10 molecules. Nothing happens in the inert soup save the very slow spontaneous chemical reactions. Increase the diversity and atomic complexity of the molecules, and more and more of the reactions among them become catalyzed by members of the system itself. As a threshold diversity is crossed, a giant web of catalyzed reactions crystallizes in a phase transition. The catalyzed reaction subgraph goes from having many disconnected tiny components to having a giant component and some smaller, isolated components. Your intuitions may now be tuned enough to guess that the giant component will contain a collectively autocatalytic subset able to form itself by catalyzed reactions from a supply of food molecules.

I have now related the central ideas about how I think life may have formed. These ideas are really very simple, if unfamiliar. Life crystallizes at a critical molecular diversity because catalytic closure itself crystallizes. These ideas, I hope, will become experimentally established parts of our new chemical creation story, our new view of our ancient roots, our new sense of the emergence of life as an expected property of the physical world.

In the computer-simulation movies we have made of this process, we can see this crystallization happening through an increase in either the diversity of molecules or the probability that any molecule catalyzes any reaction. We call these parameters M and P. As either M or P increases, at first nothing much happens in the dead soup; then suddenly it springs to life. The experiment has not been done with real chemicals yet, although I'll return to that later. But on the computer, a living system swarms into existence. Figure 3.7 shows what one of these model self-reproducing metabolisms actually looks like. As you can see, this model system is based on the continuous supply of several simple food molecules, the monomers A and B, and the four possible dimers: AA, AB, BA, and BB. From this, the system crystallizes a collectively autocatalytic, self-sustaining model metabolism with some 21 kinds of molecules. More complex autocatalytic sets have hundreds or thousands of molecular components.

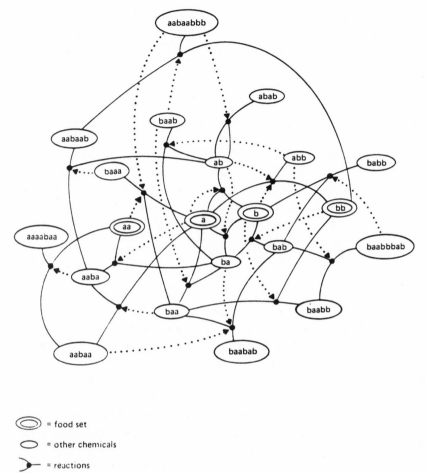

Figure 3.7 *An autocatalytic set. A typical example of a small autocatalytic set in which food molecules (a, b, aa, bb) are built up into a self-sustaining network of molecules. The reactions are represented by points connecting larger polymers to their breakdown products. Dotted lines indicate catalysis and point from the catalyst to the reaction being catalyzed.*

The same basic results are found if we use the template-matching model of catalysis. The ratio of possible reactions to polymers is so vast that eventually a giant catalyzed component and autocatalytic sets emerge. Given almost any way in which nature might determine which chemicals catalyze which reactions, a critical molecular diversity is reached at which the number of red catalyzed reactions passes a phase transition and a vast web of chemicals crystallizes in the system. This vast web is, it turns out, almost always collectively autocatalytic.

Such a system is, at minimum, self-sustaining, but such a system is very nearly self-reproducing. Suppose our collectively autocatalytic reaction system is contained within some kind of compartment. Compartmentalization must have been essential to prevent dilution of the reacting molecules. The autocatalytic system might constitute one of Alexander Oparin's coascervates, or it might create and be contained in a bilipid membrane vesicle. As the molecular constituents of the system re-create themselves, the number of copies of each kind of molecule can increase until the total has doubled. The system can then break into two coascervates, two bilipid membrane vesicles, or other compartmentalized forms. In fact, such breaking in two happens spontaneously as such systems increase in volume. Thus our autocatalytic protocell has now self-reproduced. A self-reproducing chemical system, alive by these criteria, springs into existence.

Energizing the Reactions

Now one might object that what is true for As and Bs may not be true for atoms and molecules. As Einstein said, a theory should be as simple as possible, but not too simple. One thing lacking in our model so far has been energy. As we have seen, living systems are open, nonequilibrium thermodynamic systems, sustained by a flux of matter and energy through them. As with the vastly simpler Belosov–Zhabotinski reaction, living systems maintain structures by dissipating matter and energy—in short, by eating and excreting.

The problem is this: it takes energy to create large polymers, for thermodynamics favors their breakdown into smaller constituents. A chemically realistic autocatalytic set has to obtain energy to create and sustain large molecules that may be its catalysts.

To be concrete, consider a protein with 100 amino acids linked together, or even a smaller sequence of amino acids, called a peptide. The linking of any two amino acids by a peptide bond requires energy. An easy way to see this is that the bond confines the motion of the two amino acids relative to each other. It would require some tugging to pull the amino acids apart. The tugging required is a measure of the energy of the bond. I noted earlier that almost all reactions are spontaneously reversible. This is true of a peptide bond. During its formation, a water molecule is pulled out of the reacting pair of amino acids. Thus water itself is a product of the reaction. Conversely, when a peptide bond is cleaved, a water molecule is used up. If peptides are dissolved in water, the water molecules will tend to break peptide bonds.

In a normal aqueous environment, the equilibrium ratio of cleaved amino acids to amino acid pairs (dipeptides) is about 10 to 1. But the same calculation holds for a dipeptide plus a single amino acid coming together to form a tripeptide. In an aqueous environment, the ratio of the dipeptide and amino acid to the tripeptide will be about 10 to 1 at chemical equilibrium. Note the consequence: at equilibrium, the ratio of two amino acids to the dipeptide they form is 10 to 1, and the ratio of the dipeptide plus a single amino acid to the tripeptide is also 10 to 1. Thus the ratio of single amino acids to tripeptides is not 10 to 1, but roughly 100 to 1. Similarly, at equilibrium, the ratio of amino acids to tetrapeptides is about 1,000 to 1. As the bigger polymer increases in length, its equilibrium concentration relative to the amino acids falls by a factor of about 10-fold for each increase of one amino acid in length.

The implication of the previous simple calculation is this: in an equilibrium mixture of single amino acids and various peptides up to length, say 25, the average ratio of the amino acid concentrations to that of any specific peptide of 25 amino acids would be about 1 to 10^{-25}. To be concrete, if amino acids were dissolved to the highest concentration in water that can be attained, then at equilibrium the number of copies of any specific sequence of amino acids 25 residues long would be less than one molecule in a liter of water! By contrast, the number of copies of any of the single amino acids might be on the order of 10^{20} to 10^{23}. Autocatalytic sets may use large polymers. How can high concentrations of such molecules be achieved in the face of this thermodynamic difficulty?

There are at least three fundamental ways that this vast obstacle might have been overcome. Each is remarkably simple. First, reactions can be confined to a surface rather than occurring in a volume. The reason this helps form larger polymers is simple. The rate at which a chemical reaction occurs depends on how rapidly the reaction partners collide with one another. If an enzyme is involved, the enzyme must be encountered as well. If the reaction is occurring in a volume, such as a beaker, then each molecule must diffuse in three dimensions and bump into its reaction partners. It is rather easy for molecules wandering in three dimensions to keep missing one another (recall the cartoon I described in Chapter 2). By contrast, if the molecules are confined to a very thin surface layer, such as clay or a bilipid membrane, then the search occurs in only two dimensions. It is a lot harder for the molecules to miss one another. To tune your intuition, imagine the molecules diffusing in a one-dimensional tube with a tiny diameter. Then they are bound to run into one another. In short, confining reactions to occur on surfaces strongly increases the chances of substrates hitting one another, hence enhancing the rate of formation of longer polymers.

A second simple mechanism to enhance the formation of longer polymers is to dehydrate the system. Dehydration removes water molecules, hence slowing down the cleavage of peptide bonds. In computer simulations with my colleagues Doyne Farmer, Norman Packard, and, later, Richard Bagley, we found strong evidence that even simple dehydration ought to suffice to allow real autocatalytic systems of polymers to reproduce. Our model fits the laws of chemistry and physics without straining.

Dehydration is not a cheat; it actually works. A famous reaction, called the plastein reaction, was well studied beginning almost 60 years ago. The enzyme trypsin in the stomach helps digest the proteins we eat. If trypsin is mixed with large proteins in an aqueous medium, it cleaves the proteins into smaller peptides. But if the reaction system is dehydrated, lowering the concentration of water relative to the peptides, the equilibrium shifts in favor of the synthesis of larger polymers from the small peptide fragments. Trypsin obliges by catalyzing these ligation reactions, yielding larger polymers. If these larger polymers are removed and the system is again dehydrated, trypsin obliges by synthesizing still more large polymers.

Reactions on surfaces and dehydration can be used to favor the formation of large polymers. But contemporary cells also use a more flexible and sophisticated mechanism. As cells form bonds, they obtain the needed energy by simultaneously breaking down the high-energy bonds in ubiquitous helper molecules. Adenosine triphosphate (ATP) is the most common of these. Reactions that require energy are called endergonic; those that release energy are called exergonic. Cells drive endergonic reactions by linking them to exergonic reactions.

A number of plausible candidates have been suggested for high-energy bonds that may have powered early self-reproducing metabolisms. For example, pyrophosphate, two phosphates linked together, is abundant and releases substantial energy upon cleavage. Pyrophosphate may have been a useful source of free energy to drive synthesis in early living systems. Farmer and Bagley have used computer simulations to show that model systems powered by these bonds meet plausible thermodynamic criteria and can reproduce.

What is required to link exergonic and endergonic reactions? Does some new mystery confront us beyond the achievement of catalytic closure? I think not. A problem is here, but hardly a mystery. All that is required, after all, is that the autocatalytic set include catalysts that link exergonic and endergonic reactions, so that one powers the other. The endergonic synthesis of large molecules must be coupled with the degradation of high-energy bonds supplied by food molecules or, ultimately, sunlight. But this does not seem an overwhelming obstacle.

Catalysis of such coupled reactions is not fundamentally different from other reactions: an enzyme able to bind the transition state is needed. All that is required is a sufficient diversity of molecules.

An Unrepentant Holism

This theory of life's origins is rooted in an unrepentant holism, born not of mysticism, but of mathematical necessity. A critical diversity of molecular species is necessary for life to crystallize. Simpler systems simply do not achieve catalytic closure. Life emerged whole, not piecemeal, and has remained so. Thus unlike the dominant nude RNA view of the origin of life, with its evolutionary just-so stories, we have a hope of explaining why living creatures seem to have a minimal complexity, why nothing simpler than the pleuromona can be alive.

If this view is right, we should be able to prove it. We should be able to create life anew in the fabled test tube, as though it were held by some scientist driven by Faustian dreams. Can we hope to make a new life-form? Can we brazen the face of God? Yes, I think so. And God, in his grace and simplicity, should welcome our struggles to find his laws. The ways of science are genuinely mysterious. As we shall see in Chapter 7, the hope to create collectively autocatalytic sets of molecules is linked to what may become the second era of biotechnology, promising new drugs, vaccines, and medical miracles. And the concept of catalytic closure in collectively autocatalytic sets of molecules will begin to appear as a deep feature of the laws of complexity, reemerging in our understanding of ecosystems, economic systems, and cultural systems.

Immanuel Kant, writing more than two centuries ago, saw organisms as wholes. The whole existed by means of the parts; the parts existed both because of and in order to sustain the whole. This holism has been stripped of a natural role in biology, replaced with the image of the genome as the central directing agency that commands the molecular dance. Yet an autocatalytic set of molecules is perhaps the simplest image one can have of Kant's holism. Catalytic closure ensures that the whole exists by means of the parts, and they are present both because of and in order to sustain the whole. Autocatalytic sets exhibit the emergent property of holism. If life began with collectively autocatalytic sets, they deserve awed respect, for the flowering of the biosphere rests on the creative power they unleashed on the globe—awed respect and wonder, but not mysticism.

Most important of all, if this is true, life is vastly more probable than we have supposed. Not only are we at home in the universe, but we are far more likely to share it with as yet unknown companions.

Chapter 4

Order for Free

The living world is graced with a bounty of order. Each bacterium orchestrates the synthesis and distribution of thousands of proteins and other molecules. Each cell in your body coordinates the activities of about 100,000 genes and the enzymes and other proteins they produce. Each fertilized egg unfolds through a sequence of steps into a well-formed whole called, appropriately enough, an organism. If the sole source of this order is what Jacques Monod called "chance caught on the wing," the fruit of one fortuitous accident after another and selection sifting, then we are indeed improbable. Our lapse from paradise—Copernicus to Newton in celestial mechanics, to Darwin in biology, and to Carnot and the second law of thermodynamics—leaves us spinning around an average star at the edge of a humdrum galaxy, lucky beyond reckoning to have emerged as living forms.

How different is humanity's stance, if it proves true that life crystallizes almost inevitably in sufficiently complex mixtures of molecules, that life may be an expected emergent property of matter and energy. We start to find hints of a natural home for ourselves in the cosmos.

But we have only begun to tell the story of emergent order. For spontaneous order, I hope to show you, has been as potent as natural selection in the creation of the living world. We are the children of twin sources of order, not a singular source. So far we have showed how autocatalytic sets might spring up naturally in a variegated chemical soup. We have seen that the origin of collective autocatalysis, the origin of life itself, comes because of what I call "order for free"—self-organization that arises naturally. But I believe that this order for free, which has undergirded the origin of life itself, has also undergirded the order in organisms as they have evolved and has even undergirded the very capacity to evolve itself.

If life emerged as collectively autocatalytic systems swirling in some soup, then our history only starts there. It had best not end abruptly for lack of the ability to evolve. The central motor of evolution, Darwin taught us, requires self-reproduction and heritable variation. Once these occur, natural selection will cull the fitter from the less fit. Most biologists hold that DNA or RNA as a stable store of genetic information is essential to adaptive evolution. Yet if life began with collective autocatalysis and later learned to incorporate DNA and the genetic code, we are faced with explaining how such autocatalytic sets could undergo heritable variation and natural selection without yet harboring a genome. If we required the magic of template replication and the further magic of genetic coding for proteins, the chicken-and-egg problem becomes too horrendous to contemplate. Evolution cannot proceed without these mechanisms, and we cannot have these mechanisms without evolution to tinker them together. In continuing our search for a theory of we the expected, we are led to ask this question: Is there a way that an autocatalytic set could evolve without all the complications of a genome?

My colleagues Richard Bagley and Doyne Farmer have hinted at how this might happen. We have already seen in Chapter 3 that once an autocatalytic set is enclosed in a spatial compartment of some sort—say, a coascervate or a bilipid membrane vesicle—the self-sustaining metabolic processes can actually increase the number of copies of each type of molecule in the system. In principle, when the total has doubled, the compartmentalized system can "divide" into two daughters. Self-reproduction can occur. As I noted, in experiments such compartmental systems do tend to divide spontaneously into two daughters as their volumes increase. But if daughter "cells" were always identical to the parent "cell," no heritable variation could occur.

Richard and Doyne found a natural way that variation and evolution in such systems can occur. (Richard did this work as part of his doctoral dissertation at the University of California, San Diego, with Stanley Miller as one of his examiners.) They proposed that a random, uncatalyzed reaction will occasionally occur as an autocatalytic net goes about its business. These spontaneous fluctuations will tend to give rise to molecules that are not members of the set. Such novel molecules can be thought of as a kind of penumbra of molecular species, a chemical haze surrounding the autocatalytic set. By absorbing some of these new molecular species into itself, the set would become altered. If one of these new molecules helped catalyze its own formation, it would become a full-fledged member of the network. A new loop would be added to the metabolism. Or if the molecular interloper inhibited a previously occurring reaction, then an old loop might be eliminated from the set. Either way, heritable variation was evidently possible. If the result were a more efficient network—one better able to sustain itself amid a harsh environment—then these mutations would be rewarded, the altered web crowding out its weaker competitors.

In short, there is reason to believe that autocatalytic sets can evolve without a genome. This is not the kind of evolution we are accustomed to thinking about. There is no separate DNA-like structure carrying genetic information. Biologists divide cells and organisms into the genotype (the genetic information) and the phenotype (the enzymes and other proteins, as well as the organs and morphology, that make up the body). With autocatalytic sets, there is no separation between genotype and phenotype. The system serves as its own genome. Nevertheless, the capacity to incorporate novel molecular species, and perhaps eliminate older molecular forms, promises to generate a population of self-reproducing chemical networks with different characteristics. Darwin tells us that such systems will evolve by natural selection.

In fact, such self-reproducing, compartmentalized protocells and their daughters will inevitably form a complex ecosystem. Each protocell reproduces with heritable variations; in addition, each will tend to absorb and excrete molecular species selectively in its environment, as do contemporary bacteria. In short, a molecule created in one protocell can be transported to other protocells. That molecule may promote or poison reactions in the second protocell. Not only does metabolic life begin whole and complex; but all the panoply of mutualism and competition that we think of as an ecosystem springs forth from the very beginning. The story of such ecosystems at all scales is the story not merely of evolution, but of coevolution. We have all made our worlds together for almost 4 billion years. The story of order for free continues in this molecular and organismic coevolution, as will be shown in later chapters.

But evolution requires more than simply the ability to change, to undergo heritable variation. To engage in the Darwinian saga, a living system must first be able to strike an *internal* compromise between malleability and stability. To survive in a variable environment, it must be stable, to be sure, but not so stable that it remains forever static. Nor can it be so unstable that the slightest internal chemical fluctuation causes the whole teetering structure to collapse. We have only to consider again the now familiar concepts of deterministic chaos to appreciate the problem. Recall the famous butterfly in Rio, whose energetic wing flapping, or even languid stirring, can alter the weather in Chicago. In chaotic systems, tiny changes in initial conditions can lead to profound disturbances. From what we have said so far, there is no reason to believe that our autocatalytic sets would not be hypersensitive, chaotic, doomed from the start. A tiny change in the concentrations of the internal metabolism because some molecule from a neighboring cell is absorbed might be amplified so mightily that the network would fly apart. The autocatalytic sets I am proposing would have had to coordinate the behaviors of some thousands of molecules. The chaos that could potentially flourish in systems of this complexity boggles the mind.

The potential for chaos is not merely theoretical. Other molecules can bind to the enzymes in our own cells, inhibiting or increasing their activity. Enzymes can be "turned on" or "turned off" by other molecules in the reaction network. It is now well known that in most cells, such molecular feedback can give rise to complex chemical oscillations in time and space. The potential for chaos is real.

If we are to believe that life began when molecules spontaneously joined to form autocatalytic metabolisms, we will have to find a source of molecular order, a source of the fundamental internal homeostasis that buffers cells against perturbations, a compromise that would allow the protocell networks to undergo slight fluctuations without collapsing. How, without a genome, would such order arise? It must somehow emerge from the collective dynamics of the network, the coordinated behavior of the coupled molecules. It must be another case of order for free. As we are about to see, astonishingly simple rules, or constraints, suffice to ensure that unexpected and profound dynamical order emerges spontaneously.

The Wellsprings of Homeostasis

Allow me a simple, highly useful, idealization. Let us imagine that each enzyme has only two states of activity—on or off, and can switch between them. So at each moment, each enzyme is either active or inactive. This idealization, like all idealizations, is literally false. In reality, enzymes show graded catalytic activities. Most simply, the rate of a reaction depends on enzyme and substrate concentrations. Nevertheless, inhibition or activation of enzymes by molecules binding to sites on the enzyme or changing the enzyme in other ways is common and is often associated with a sharp change in enzyme activity. In addition, allow me to think of the substrates or products of reactions as either present or absent. This, too, is literally false. But often the concentrations of substrates and products in complex reaction networks can change very swiftly from high to low. The "on–off" "present–absent" idealization is very useful, for we are going to consider networks with thousands of model enzymes, substrates, and products.

The point in using idealizations in science is that they help capture the main issues. Later one must show that the issues so captured are not altered by removing the idealizations. Thus in physics, analysis of the gas laws was based on models of gas molecules as hard elastic spheres. The idealization captured the main features necessary to create statistical mechanics. In Chapter 3, we presented molecules and their reactions as buttons and threads. Now let us change metaphors and think of a metabolic network of enzymes, substrates, and products as a network of lightbulbs connected by wires, an electrical circuit. A molecule catalyzing the formation of another molecule can be thought of as one bulb turning on another. But molecules can also inhibit each other's formation. Think of this as one bulb turning another bulb off.

One way to get such a network to behave in an orderly manner would be to design it with great care and craft. But we have proposed that autocatalytic metabolisms arose in the primal waters spontaneously, built from a random conglomeration of whatever happened to be around. One would think that such a haphazard concoction of thousands of molecular species would most likely behave in a manner that was disorderly and unstable. In fact, the opposite is true: order arises spontaneously, order for free. To return to our metaphor, although we wire our bulbs together at random, they do not necessarily blink on and off randomly like the twinkling lights of a vast forest of berserk Christmas trees. Given the right conditions, they settle into coherent, repeating patterns.

To see why order emerges spontaneously, I have to introduce some of the concepts mathematicians use to think about dynamical systems. If we think of our autocatalytic set as an electrical network, then it can assume a vast number of possible states. All the bulbs might be off, all might be on, and in between these two extremes can be myriad combinations. Imagine a network that consists of 100 nodes, each of which can be in one of two possible states, either on or off; the number of possible configurations is 2^{100}. For our autocatalytic metabolism, with perhaps 1,000 kinds of molecules, the number of possibilities is even vaster: $2^{1,000}$. This range of possible behaviors is called a state space. We can think of it as the mathematical universe in which the system is free to roam.

To make these notions concrete, consider a simple network consisting of just three light bulbs—1, 2, and 3—each of which receives "inputs" from the other two. (Figure 4.1*a*). The arrows show which way

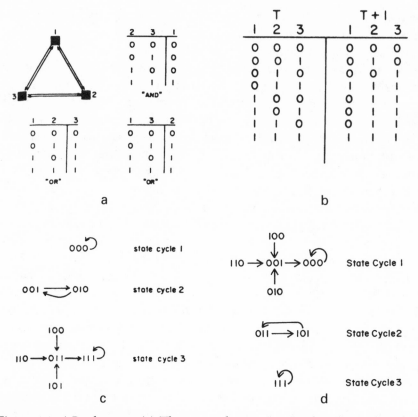

Figure 4.1 *A Boolean net.* (a) *The wiring diagram in a Boolean network with three binary elements, each an input to the other two.* (b) *The Boolean rules of (a) rewritten to show for all (2³) = 8 states at time* T, *the activity assumed by each element at the next moment,* T + 1. *Read from left to right, this figure shows the successor state for each state.* (c) *The state transition graph, or "behavior field," of the autonomous Boolean network of (a) and (b), obtained by showing state transitions to successor states connected by arrows.* (d) *Effects of mutating the rule of element 2 from* OR *to* AND.

the signals flow; thus arrows point to bulb 1 from bulbs 2 and 3, signifying that bulb 1 receives inputs from bulbs 2 and 3.

In addition to denoting the wiring diagram, we need to know how each lightbulb responds to the signals it receives. Since each bulb can have only two values, on and off, which we can represent as 1 and 0, then it is easy to see that there are four possible input patterns it can receive from its two neighbors. Both inputs can be off (00), one or the other input can be on (01 or 10), or both inputs can be on (11). Using this information, we can construct a rule table specifying whether each bulb will be active (1) or inactive (0) for each of these four possible signals. For example, bulb 1 might be active only if both of its inputs were active the moment before. In the language of Boolean algebra (named in honor of George Boole, the inventor of mathematical logic in the nineteenth century), bulb 1 is an AND gate: bulbs 2 *and* 3 must be active before it will light. Or we could choose instead for the bulb to be governed by the Boolean OR function: bulb 1 will be active the next moment if bulb 2 or bulb 3 or both were active the moment before.

To complete the specification of what I will now call a Boolean network, I will assign to each lightbulb one of the possible Boolean functions. Say I assign the AND function to bulb 1 and the OR function to bulbs 2 and 3 (Figure 4.1*a*). At each tick of the clock, each bulb examines the activities of its two inputs and adopts the state 1 or 0 specified by its Boolean function. The result is a kaleidoscopic blinking as pattern after pattern unfolds.

Figure 4.1*b* shows the eight possible states that the network can assume, from (000) to (111). Read along vertical columns, the right half of Figure 4.1*b* specifies the Boolean rule governing each lightbulb. But read from left to right, Figure 4.1*b* shows, for each current state at time T, the next state of the entire network one moment later, at $T + 1$, when all lightbulbs simultaneously adopt their new activities, 1 or 0.

Now we are in a position to begin to understand the behavior of this little network. As we can see, the system can be in a finite number of states, here eight. If started in one state, over time the system will flow through some sequence of states. This sequence is called a trajectory (Figure 4.1*c*). Since there is a finite number of states, the system must eventually hit a state it has previously encountered. Then the trajectory will repeat. Since the system is deterministic, it will cycle forever around a recurrent loop of states called a state cycle.

Depending on the initial state in which we start our network—the pattern of on and off bulbs—it will follow various trajectories, falling at some point into an ever repeating state cycle (Figure 4.1*c*). The simplest possible behavior would occur if the network fell immediately into a state cycle consisting of a single pattern of 1s and 0s. A system started in such a state never changes; it is said to be stuck in a cycle of length 1. Alternatively, the length of the state cycle could conceivably be the total number of states in state space. A system caught in such a cycle will repeat, one after another, every pattern it is capable of displaying. For our three-bulb system, this would result in a steady twinkling as the system passed through its eight possible states. Since the number of states is so small, we could soon detect the pattern of its blinking. Now imagine a larger network, with 1,000 bulbs and thus $2^{1,000}$ possible states. If the network were on a state cycle passing through every one of this hyperastronomical number of states, and if it took a mere *trillionth of a second* per state transition, we could never in the lifetime of the universe see the system complete its orbit.

So the first thing to appreciate about Boolean networks is this: any such network will settle down to a state cycle, but the number of states on such a recurrent pattern might be tiny—as small as a single steady state—or so hyperastronomical that numbers are meaningless. If a system falls into a small state cycle, it will behave in an orderly manner. But if the state cycle is too vast, the system will behave in a manner that is essentially unpredictable. The state spaces through which molecular networks with only a few thousand kinds of molecules can roam are beyond our common reckoning. For our autocatalytic networks to be orderly, they must avoid veering off on seemingly endless tangents and must settle down into small state cycles—a repertoire of stable behaviors.

To gain insight into how likely it is that autocatalytic sets would be stable enough to endure, we must ask these questions: How does one make orderly networks with short state cycles? Is the creation of tiny state cycles difficult, meaning that it is something of a miracle that stable autocatalytic metabolisms emerged? Or does it happen naturally? Is it part of order for free?

To answer these questions we need to understand the concept of an attractor. More than one trajectory can flow into the same state cycle. Start a network with any of these different initial patterns and, after churning through a sequence of states, it will settle into the same state cycle, the same pattern of blinking. In the language of dynamical systems, the state cycle is an attractor and the collection of trajectories that flow into it is called the basin of attraction. We can roughly think of an attractor as a lake, and the basin of attraction as the water drainage flowing into that lake.

Just as a mountainous region may harbor many lakes, a Boolean network may harbor many state cycles, each draining its own basin of attraction. The little network in Figures 4.1*a–c* has three state cycles. The first state cycle has the single steady state (000), which drains no basin of trajectories. It is an isolated steady state. It can be reached only if we start the network there. The second state cycle has two states, (001) and (010). The network oscillates between these two. No other states drain into this attractor. Launch the network with one of these two patterns and it will remain in the cycle, blinking back and forth between the two states. The third state cycle consists of the steady state (111). This attractor lies in a basin of attraction draining four other states. Start the network with any one of these patterns and it will quickly flow to the steady state and freeze up, displaying three lighted bulbs.

Under the right conditions, these attractors can be the source of order in large dynamical systems. Since the system follows trajectories that inevitably flow into attractors, tiny attractors will "trap" the system into tiny subregions of its state space. Among the vast range of possible behaviors, the system settles into an orderly few. The attractors, if small, create order. Indeed, tiny attractors are a prerequisite for the order for free that we are seeking.

But tiny attractors are not enough. For a dynamical system, such as an autocatalytic net, to be orderly, it must exhibit homeostasis; that is, it must be resistant to small perturbations. Attractors are the ultimate source of homeostasis as well, ensuring that a system is stable. In large networks, any state cycle typically drains an enormous basin; many states flow into the attractor. Moreover, the states within that basin can be very similar to the states on the state cycle to which they drain. Why is this important? Suppose we arbitrarily choose a single lightbulb and flip it to the opposite state. All or most such perturbations leave the system in the same basin of attraction. So the system will return to the same state cycle from which it was perturbed! That is the essence of homeostatic stability. State cycle 3 in Figure 4.1c is stable in this way; if the network is in this basin, flipping the activity of any single lightbulb will have no long-term impact on its behavior, for the system will return to the same state cycle.

But homeostatic stability does not always arise. State cycle 1, by contrast, is an isolated steady state and is unstable to the slightest perturbation. After any such flip, the system is shoved into a different basin of attraction. It can't come home again. If the network had the property that all attractors were unstable in this way, we can imagine that slight perturbations (the flapping of the butterfly's wings) would persistently bump the system out of attractors and send it veering off on an endless, never repeating journey through state space. The system would be chaotic.

If we are to believe that life began with the spontaneous generation of autocatalytic nets, then we had better hope that they were homeostatic. Is it natural that certain kinds of large networks will exhibit homeostasis? Is homeostasis hard to create, making the emergence of stable networks vastly unlikely? Or can it, too, be part of order for free?

What we need are laws describing which kinds of networks are likely to be orderly and which are likely to succumb to chaos. Any Boolean network has attractors, each draining some basin of attraction, but the state spaces of networks with thousands of kinds of molecular species are hyperastronomical. If we change metaphors and think of each state cycle as a galaxy in space, how many attractor galaxies are there spread across the mega-mega parsecs of network state spaces? With umpteen gadzillions of states in state space, one might have gadzillions of attractors. If there are vast numbers of attractors, and the system might be located on any one of them, that does not sound like order.

Collectively autocatalytic sets presumably evolved, and contemporary organisms do evolve, by mutations that permanently change the functional connections among the molecular species in the system. Will such permanent mutational changes cause an autocatalytic system to collapse into chaotic twinkling through its space of molecules, poisoning its own capacity to catalyze its own reproduction? Will minor mutational variations typically cause catastrophic changes? In the language of Boolean networks, another way to perturb a network is to permanently "mutate" its wiring diagram, changing the inputs or the Boolean function governing when a bulb is on or off. In Figure 4.1d, I show the result of changing the rule governing lightbulb 2 from OR to AND. As you can see, this causes the network to assume a new dynamical form. Some state cycles remain, but others are changed. New basins of attraction will steer the network into different patterns.

Darwin supposed that living systems evolve by mutations that cause small modifications in the properties of the organism. Is this graceful property of minor changes hard to achieve? Or is it, too, part of order for free? A pure Darwinist might argue that this kind of graceful stability could arise only after a series of evolutionary experiments, yet this begs the question. We are trying to explain the origin of the very ability to evolve! However life started, with nude replicating RNA molecules or with collectively autocatalytic sets, this stability cannot be imposed from outside by natural selection. It must arise from within as a condition of evolution itself.

All these properties we need, I believe, all the order we require, arise spontaneously. We next must show how order for free supplies the small ordered attractors we need, the homeostasis we need, and the graceful stability we need. Order for free, utterly natural, if previously mostly unknown, will change our view of life.

The Requirements for Order

We have seen that Boolean networks can exhibit profound order, but Boolean networks can also exhibit profound chaos. Consequently, we seek the conditions under which orderly dynamics can emerge in such systems. I will now present the results of about 30 years of work.

The main results are simple to summarize: two features of the way networks are constructed can control whether they are in an ordered regime, a chaotic regime, or a phase transition regime between these, "on the edge of chaos." One feature is simply how many "inputs" control any lightbulb. If each bulb is controlled by only one or two other lightbulbs, if the network is "sparsely connected," then the system exhibits stunning order. If each bulb is controlled by many other lightbulbs, then the network is chaotic. So "tuning" the connectivity of a network tunes whether one finds order or chaos. The second feature that controls the emergence of order or chaos is simple biases in the control rules themselves. Some control rules, the AND and OR Boolean functions we talked about, tend to create orderly dynamics. Other control rules create chaos.

The way I and others have done this work is pretty straightforward. One way to ask what kinds of lightbulb networks exhibit order of chaos is to construct very specific networks and study them. But this would leave us with a vast number of very specific networks to study—another of our hyperastronomical numbers, big beyond meaning. The approach I have taken asks whether networks of certain *general kinds* exhibit order or chaos. To answer this question, the natural approach is to carefully define the "kind" of networks in question, and then use computers to simulate large numbers of networks drawn at random from the pool. Then, like a pollster, we can build up a portrait of the typical, or generic, behaviors of members of the class.

We might, for example, study the pool of networks with 1,000 bulbs (we'll call this variable N) and 20 inputs per bulb (the variable K). Given $N = 1000$ and $K = 20$, a vast ensemble of networks can be built. We sample this ensemble by randomly assigning to each of the 1,000 bulbs 20 inputs and, again at random, one of the possible Boolean functions. Then we can study the network's behavior, counting the number of attractors, the lengths of attractors, the stability of attractors to perturbations and mutations, and so forth. Throwing the dice again, we can randomly wire another network with the same general characteristics and study its behavior. Sample by sample, we build up a portrait of a family of Boolean nets, and then we change the values of N and K and build up another portrait.

After years of such experiments, networks with various parameters become as familiar as old friends. Consider networks in which each lightbulb receives input from only one other. In these $K = 1$ networks, nothing very interesting happens. They quickly fall into very short state cycles, so short that they often consist of but a single state, a single pattern of illumination. Launch such a $K = 1$ network and it freezes up, saying the same thing over and over for all time.

At the other end of the scale, consider networks in which $K = N$, meaning that each lightbulb receives an input from all lightbulbs, including itself. One quickly discovers that the length of the networks' state cycles is the square root of the number of states. Consider the implications. For a network with only 200 binary variables—bulbs that can be on or off—there are 2^{200} or 10^{60} possible states. The length of the state cycles is thus on the order of 10^{30} states. Start the network with some arbitrary pattern of on-bulbs and off-bulbs, 1s and 0s, and it will be pulled by an attractor into a repeating cycle, but a cycle so long as to be all but fathomless. Suppose the network took a millionth of a second to pass from state to state. Then the little network would require 10^{30} millionths of a second to traverse its state cycle. This is equal to billions of times the 15-billion-year history of the universe! So we could never actually observe the fact that the system had "settled" onto its state cycle attractor! We could never tell from the twinkling patterns of the lightbulbs that the network was not just wandering randomly around in its entire state space!

I hope this gives you pause. We are searching for laws that suffice to yield orderly dynamics. Our Boolean networks are nonequilibrium, open thermodynamic systems. Since a little network with only 200 light-bulbs can twinkle away for an eternity without repeating a pattern, order is in no way automatic in nonequilibrium, open thermodynamic systems.

Such $K = N$ networks do show signs of order, however. The number of attractors in a network, the number of lakes, is only N/e, where e is the basis of the natural logarithms, 2.71828. So a $K = N$ network with 100,000 binary variables would harbor about 37,000 of these attractors. Of course, 37,000 is a big number, but very very much smaller than $2^{100,000}$, the size of its state space.

Suppose, then, that we perturb the network, flipping a bulb from off to on, or vice versa. In $N = K$ networks, we get an extreme version of the butterfly effect. Flip a bit, and the system almost certainly falls under the sway of another attractor. But since there are 37,000 attractors with lengths up to $10^{15,000}$ states, the tiny fluctuation will utterly change the future evolution of the system. $K = N$ networks are massively chaotic. No order for free in this family.

Even worse, try evolving such a network by randomly swapping the Boolean rule of some lightbulb. You will alter half the state transitions in the network and scatter all the old basins of attraction and state cycles to the dustbin of network history. Small changes here cause massive changes in behavior. There are no graceful minor heritable variations for selection to act on in this family.

Most Boolean networks are chaotic, and they are graceless with respect to minor mutations. Even networks in which K is much less than N, $K = 4$ or $K = 5$, exhibit unpredictable, chaotic behavior similar to that seen for $K = N$ networks.

Whence cometh the order? The order arises, sudden and stunning, in $K = 2$ networks. For these well-behaved networks, the length of state cycles is not the square root of the number of states, but, roughly, the square root of the number of binary variables. Let's pause to translate this as clearly as we can. Think of a randomly constructed Boolean network with $N = 100,000$ lightbulbs, each receiving $K = 2$ inputs. The "wiring diagram" would look like a madhatterly scrambled jumble, an impenetrable jungle. Each lightbulb has also been assigned at random a Boolean function. The logic is, therefore, a similar mad scramble, haphazardly assembled, mere junk. The system has $2^{100,000}$ or $10^{30,000}$ states—megaparsecs of possibilities—and what happens? The massive network quickly and meekly settles down and cycles among the square root of 100,000 states, a mere 317.

I hope this blows your socks off. Mine have never recovered since I discovered this almost three decades ago. Here is, forgive me, stunning order. At a millionth of a second per state transition, a network, randomly assembled, unguided by any intelligence, would cycle through its attractor in 317-millionths of a second. This is a lot less than billions of times the history of the universe. Three hundred seventeen states? To see what this means in another way, one can ask how tiny a fraction of the entire state space the network squeezes itself into. A mere 317 states compared with the entire state space is an extremely tiny fraction of that state space, about 1 divided by $10^{29,998}$!

We seek order without careful crafting. Recall our discussion in Chapter 1 of closed thermodynamic systems in which the gas molecules diffuse from improbable configurations—clumped in one corner or spread parallel to one face of a box—toward homogeneous configurations. The improbable configurations constituted order. Here, in this class of open thermodynamic systems, the spontaneous dynamics drive the system into an infinitesimal corner of its state space and hold it there, quivering for an eternity. Order for free.

Order expresses itself in these networks in diverse ways. Nearby states converge in state space. In other words, two similar initial patterns will likely lie in the same basin of attraction, hence driving the system to the same attractor. Thus such systems do not show sensitivity to initial conditions; they are not chaotic. The consequence is the homeostasis we seek. Once such a network is on an attractor, it will return to the same attractor with very high probability if it is perturbed. Homeostasis is free in this neck of the network woods.

For the same reason, these networks can undergo a mutation that alters wiring or logic without veering into randomness. Most small mutations cause our hoped-for small, graceful alteration in the behavior of the network. Basins and attractors change only slightly. Such systems evolve readily. So selection does not have to struggle to achieve evolvability.

Finally, these networks are not *too* orderly. Unlike the $N = 1$ network, they are not frozen like a rock, but are capable of complex behaviors.

Our intuitions about the requirements for order have, I contend, been wrong for millennia. We do not need careful construction; we do not require crafting. We require only that extremely complex webs of interacting elements are sparsely coupled.

As I show in my book *The Origins of Order: Self-Organization and Selection in Evolution,* there are ways to tune networks in which K is greater than 2 so that they are also orderly, not chaotic. My colleagues Bernard Derrida and Gerard Weisbuch, both solid-state physicists at the Ecole Normale Supérieure in Paris, have shown that a variable called P can be tweaked to make a chaotic network become orderly.

The P parameter is very simple. Figure 4.2 shows three Boolean functions, each with four inputs. In each, the response of the regulated light-bulb must be specified for each of the 16 possible states of the four

A B C D	E		A B C D	E		A B C D	E
0 0 0 0	0		0 0 0 0	0		0 0 0 0	1
0 0 0 1	1		0 0 0 1	0		0 0 0 1	1
0 0 1 0	0		0 0 1 0	0		0 0 1 0	1
0 0 1 1	1		0 0 1 1	0		0 0 1 1	0
0 1 0 0	0		0 1 0 0	0		0 1 0 0	1
0 1 0 1	1		0 1 0 1	0		0 1 0 1	1
0 1 1 0	1		0 1 1 0	0		0 1 1 0	1
0 1 1 1	0		0 1 1 1	0		0 1 1 1	1
1 0 0 0	1		1 0 0 0	1		1 0 0 0	1
1 0 0 1	0		1 0 0 1	0		1 0 0 1	1
1 0 1 0	0		1 0 1 0	0		1 0 1 0	1
1 0 1 1	1		1 0 1 1	0		1 0 1 1	1
1 1 0 0	0		1 1 0 0	0		1 1 0 0	1
1 1 0 1	0		1 1 0 1	0		1 1 0 1	1
1 1 1 0	1		1 1 1 0	0		1 1 1 0	1
1 1 1 1	1		1 1 1 1	0		1 1 1 1	1
a			b			c	

Figure 4.2 *Tinkering with the* P *parameter.* (a) *A Boolean function of four inputs, in which eight of the 16 input configurations yield a 0 response, while eight of the 16 yield a 1 response.* P $= 8/16 = .50.$ (b) *The response is 0 for 15 of the 16 possible input configurations.* P $= 15/16 = .9375.$ (c) *Fifteen of the 16 possible input configurations yield a 1 response.* P $= 15/16 = .9375.$

input lightbulbs, from (0000) to (1111). For the Boolean function shown in Figure 4.2*a*, half the responses of the regulated lightbulb are 1, and the other half are 0. For the Boolean function shown in Figure 4.2*b*, 15 of the responses are 0, and only a single input pattern gets a 1 response from the regulated bulb. The Boolean function in Figure 4.2*c* is similar to that in Figure 4.2*b*, except that the preferred output response is 1, not 0. Fifteen of the 16 input patterns lead to a 1 response. *P* is just a parameter that measures the bias away from half 1 and half 0 responses in a Boolean function. So *P* for the Boolean function in Figure 4.2*a* is 0.5, while *P* for the Boolean function in Figure 4.2*b* is 15/16, or 0.9375, and *P* for the Boolean function in Figure 4.2*c* is also 15/16, or 0.9375.

What Bernard and Gerard showed is, after the fact, pretty intuitive. If different networks are built with increasing *P* biases, starting from the no-bias value of 0.5 to the maximum value of 1.0, networks with *P* = 0.5 or only slightly greater than 0.5 are chaotic and networks with *P* near 1.0 are orderly. This can be easily seen in the limit when their *P* parameter is 1.0. Then the bulbs in the network are of only two types. One type responds with a 0 to any input pattern; the other responds with a 1 to any input pattern. So if you start the network in any state at all, the 0-type bulbs respond with 0, the 1-type bulbs respond with 1, and the network freezes into the corresponding pattern of 0 and 1 values and remains at that steady state forever. So when the *P* parameter is maximum, networks are in an ordered regime. When the *P* parameter is 0.5, networks with many inputs per lightbulb are in a chaotic regime, twinkling away for an eternity. And, for any network, Bernard and Gerard showed that there is a critical value of *P* where the network will switch from chaotic to ordered. This is the edge of chaos, to which we will return in a moment.

The summary is this: two parameters suffice to govern whether random Boolean lightbulb networks are ordered or chaotic. Sparsely connected networks exhibit internal order; densely connected ones veer into chaos; and networks with a single connection per element freeze into mindlessly dull behavior. But density is not the only factor. If networks have dense connections, tuning the *P* bias parameter drives networks from the chaotic regime to the ordered regime.

These rules apply to networks of all sorts. In Chapter 5, I will show that the genome itself can be thought of as a network in the ordered regime. Thus some of the orderliness of the cell, long attributed to the honing of Darwinian evolution, seems likely instead to arise from the dynamics of the genomic network—another example of order for free. Again, I hope to persuade you that selection is not the sole source of order in the living world. The powerful spontaneous order we are discussing now is likely to have played a role not only in the emergence of stable autocatalytic sets, but in the later evolution of life.

The Edge of Chaos

Living systems, from the collectively autocatalytic protocells we discussed in Chapter 3 to cells in your body to whole organisms, surely must have networks that behave stably, that exhibit homeostasis and graceful minor modifications when mutated. But cells and organisms must not be too rigid in their behavior if they are to cope with a complex environment. The protocell had best be able to respond to novel molecules floating its way. The *E. coli* in your intestine copes with an enormous variety of molecules by sending internal molecular signals cascading among its enzymes and genes, triggering a variety of changes in enzyme and gene activities bent on protecting the cell from toxins, metabolizing food, or, occasionally, exchanging DNA with other cells.

How do cell networks achieve both stability and flexibility? The new and very interesting hypothesis is that networks may accomplish this by achieving a kind of poised state balanced on the edge of chaos.

We have already seen hints of an axis running from orderly behavior to chaotic behavior in our lightbulb models. Sparsely connected networks, with $K = 1$ or $K = 2$, spontaneously exhibit powerful order. Networks with higher numbers of inputs per lightbulb, $K = 4$ or more, show chaotic behavior. So tuning the number of inputs per lightbulb—hence the density of the web of connections among the bulbs—from low to high tunes networks from orderly to chaotic behavior. In addition, we saw that adjusting the P bias parameter from 0.5 to 1.0 also tunes whether networks are in a chaotic or an ordered regime.

We should not be too surprised if some kind of sharp change in behavior, some kind of phase transition from order to chaos, occurred along this axis. In fact, in Chapter 3 we saw such a sharp change in behavior in our toy model of the origin of life. Recall that we were connecting buttons with threads and found that the size of the largest connected cluster suddenly jumped from small to huge when the ratio of threads to buttons passed the magic value of 0.5. Below that value, only small connected clusters of buttons existed. Above that value, a giant component composed of most of the buttons emerged. This is a phase transition.

A very similar kind of phase transition occurs in our lightbulb network models. Once again, a giant cluster of connected elements will appear. But the connected cluster will not be buttons; it will be a giant cluster of lightbulbs, each of which is frozen into a fixed activity, 1 or 0. If this giant frozen component forms, the network of bulbs is in the ordered regime. If it does not form, the network is in the chaotic regime. Just between, just near this phase transition, just at the edge of chaos, the most complex behaviors can occur—orderly enough to ensure stability, yet full of flexibility and surprise. Indeed, this is what we mean by complexity.

One way to visualize what is happening in random lightbulb networks is to make a mental movie. Picture starting the network in some initial state. As the network passes along its trajectory toward, then around, its state cycle, two kinds of behavior might be seen at any lightbulb. That lightbulb might twinkle on and off in some more or less complex pattern, or the lightbulb might settle to a fixed activity, always on or always off. Let's envision two colors to distinguish these two behaviors: color lightbulbs that are twinkling on and off green, and color those that are fixed on or fixed off red.

Now consider a network in the chaotic regime, say an $N = 1,000$, $K = 20$ network. Almost all the lightbulbs are twinkling on and off; hence they are colored green. Perhaps a few bulbs or small clusters of bulbs are fixed on or fixed off; hence these are colored red. In short, only tiny clusters of frozen red bulbs exist in a vast sea of twinkling green bulbs. So a network in the chaotic regime has a vast sea of twinkling green lightbulbs and may have a few islands of frozen red bulbs.

Alternatively, suppose we simulate a lightbulb network in the ordered regime, say $N = 100,000$ and $K = 2$, a vast tangle of a network with a complexity equal to your genome or to a very large autocatalytic set. Start the network in an initial state and follow it along its trajectory toward a state cycle, then around the state cycle. At first, most of the lightbulbs are twinkling on and off, and are colored green. But as the network converges onto its state cycle, then orbits the cycle, more and more of the lightbulbs settle into fixed states of activity, frozen on or frozen off. So most of the lightbulbs are now colored red.

And now the magic. If you think of all the red lightbulbs, and ask whether they are connected to one another, just as we asked if the buttons were connected to one another by threads, you will find that the frozen red lightbulbs form a vast giant cluster of interconnected lightbulbs! A giant frozen component of lightbulbs, each frozen into either the on or into the off state, exists in Boolean networks in the ordered regime.

Of course, not all the lightbulbs in our $N = 100,000$, $K = 2$ network need be frozen; typically, small and large clusters of connected lightbulbs continue to twinkle on and off. These twinkling clusters are colored green. It is just the twinkling patterns of the clusters of connected green lightbulbs that constitute the cycling behavior of Boolean networks in the ordered regime. The lightbulbs in the giant frozen cluster of red lightbulbs do not twinkle at all.

If we looked into a typical network with $N = 100,000$ and $K = 2$, we would see a further important detail. The clusters of twinkling green lightbulbs are not themselves all interconnected. Instead, they form independent twinkling clusters, like twinkling green islands in a vast sea of frozen red lightbulbs.

So a Boolean network in the chaotic regime, as presented earlier, has a sea of ever-changing green lightbulbs twinkling on and off, with perhaps a few clusters of red lightbulbs that are frozen on or off. In contrast, a Boolean network in the ordered regime has a vast giant cluster of red lightbulbs that are frozen either on or off, a giant red cluster, with isolated islands of twinkling green bulbs. Your antennae should quiver. The phase transition from order to chaos, as parameters such as the number of inputs per lightbulb, K, or the bias parameter, P, are tuned, occurs when the giant frozen red cluster forms, leaving isolated twinkling green islands behind.

A particularly easy way to see this is to make a very simple Boolean network model on a square lattice. Here each lightbulb is connected to its four neighbors: north, south, east, and west. Each lightbulb is controlled by a Boolean function that tells it how to turn on or off depending on the current activities of its four inputs. Figure 4.3 shows such a lattice network, studied by Derrida and Weisbuch. They tuned the P bias parameter close enough to 1.0 so that the network is in the ordered regime, let the network settle into its state cycle, and then recorded the cycling period of each lightbulb. A lightbulb with a cycling period of 1 is therefore either frozen on or frozen off. In our mental picture, any such lightbulb should be colored red. Other lightbulbs are twinkling; hence these should be colored green. As Figure 4.3 shows, the period-1 frozen lightbulbs form a giant connected component that spreads across the entire lattice, leaving behind a few small and large twinkling clusters.

With Figure 4.3 in view, it is easy to explain the sensitivity to changes in initial conditions in chaotic networks and the lack of sensitivity to such perturbations in ordered networks. If a single lightbulb is flipped, one can follow the cascading changes radiating from that perturbation. In the ordered regime, such as in Figure 4.3, those rippling changes cannot penetrate the period-1 frozen red component. The giant frozen component is rather like a gigantic wall of constancy blocking off the

```
  8   8   1   1228228228228228228228228   1   1   1   1   1   1   1   1   1   1   1   1   1
  8   8   1   1   1   1   1228228228228   1   1   1   1   1   1   1   1   1   1   1   1   1
  8   8   845645645645622822822822822822822   1   1   1   1   1   10  10  10   1   1   1   1
  1   8   1   1228228228228   1   1   1   1   1   1   1   10  10  10   1   1   1   1
  1   1   1228228228228228228228228   1   1   1   1   1   1   1   1   1   1   1   1   1
  1   1   1   1   1228228228228228228228228   1   1   1   1   1   1   1   1   1   1   4   4
  1   1   1   1   1   1   1   1228228228228   1   1   1   1   1   1   1   1   1   1   1   1
  1   1   1   6   1   1228228228228   1   1   1   4   1   1   1   1   1   1   1   1   1   1
  1   4   1   6   6   6   1   1228228228228228228228   4   1   4   1   1   1   1   1   1   1
  1   4   1   1   6   6   6228228228228   1   1   1   4   1   4   1   1   1   1   1   1   1
  4   4   1   6   6   6   6   6228228228   1   1   1   1   1   1   1   1   1   1   4   4
  1   4   12  6   6   6   1228228228228   1   1   1   1   8   8   8   1   1   1   4
220   1   1   1   1   1   1   1   1228228228   1   1   1   1   8   8   8   8   1   1220
220220  1   1   1   1   1   1   1228228228228   1   1   1   1   8   8   4   8   1   1   1
220220  1   1   1   1   1   1   1228228   1   1   1   1   1   1   1   1220110   1
  1220110110   1   1   1   1   1228228   1   1   1   1   1   1   1   20  20110110
  1110110110   1   1   4   1   1228   1   1   2   4   1   1   1   1   1   20  20110110
110110110110110   1   4   1   1   1   1   1   2   4   1   1   1   1   20  20  20  20   1110
110110110  22   1   1   1   1   1   1   4   4228   1   1   1   20  20  20  20  20  20  20110
110110   1   1   1   1   1   1   1   1228   1   4   1   20  20  20  20  20  20  20110
110   22  22  22  22   1   1228228   1   1228228   1   4   4   1   1   1   4   20   2   22
 22   88  22  22   1   1   1   1228   1228228228   1   1   1   1   1   1   1   20   2   1
  1   88   1   1   1228228228228228228228228228   1   1   1   1   1   1   4   4   4   1
  1   8   1   1228228228228228228228228228   1   1   1   1   1   1   1   1   1   1
```

Figure 4.3 *Order for free. In this two-dimensional lattice, each site (lightbulb) is coupled to its four neighbors and governed by a Boolean function. When* P, *the bias in favor of a 1 or 0 response by any single variable, is increased above a critical value,* Pc, *percolation of a frozen component of lightbulbs, each fixed at 1 or 0, spans across the lattice and leaves isolated islands of twinkling lightbulbs free to vary between 1 and 0 values. The number at each point represents the cycling period of each lightbulb. Thus sites with 1 correspond to red lightbulbs frozen in either the on or the off state. Sites with numbers greater than 1 are green, twinkling on and off and forming isolated "unfrozen" islands in the sea of frozen sites. (The two-dimensional lattice is bent into a donut, or torus, by "gluing" the top edge to the bottom edge, and the left edge to the right edge. Therefore, all lightbulbs have four neighbors.)*

twinkling islands from one another. Perturbations can cascade within each twinkling island, but rarely propagate any further. That is fundamentally why our lightbulb networks in the ordered regime exhibit homeostasis.

But in the chaotic regime, a vast sea of twinkling lightbulbs extends across the entire network. If any such lightbulb is flipped, the consequences cascade throughout that unfrozen sea, creating massive changes in the activity patterns of the lightbulbs. So chaotic systems show massive sensitivity to small perturbations. Here, in our Boolean networks in the chaotic regime, is the butterfly effect. Flap your wings, oh butterfly, moth, or starling, briskly or languidly, and you will change the behavior of lightbulbs from Alaska to Florida.

Protocells and your cells, early life and all life, must be capable of orderly yet flexible behavior. What kinds of networks of interacting molecules, or interacting anything, are naturally capable of such ordered yet flexible behavior? Is such behavior hard to achieve? Or might it, too, be part of order for free? Now that we begin to understand order and chaos in networks coupling hundreds of thousands of lightbulbs, an answer, crisp and lovely, perhaps even true, suggests itself: perhaps networks just at the phase transition, just poised between order and chaos, are best able to carry out ordered yet flexible behaviors.

Here is a beautiful working hypothesis. Chris Langton at the Santa Fe Institute has stressed this important possibility more than any other scientist, and we can see intuitively that the edge of chaos might be an attractive regime to coordinate complex behavior. Suppose one wished to have a lattice of lightbulbs that coordinated the activities of two widely separated lightbulb sites on the lattice; suppose the lattice were in the chaotic regime, with an unfrozen sea. Then minor perturbations of the activities of one lightbulb would unleash cascades of alterations in activities, which would propagate throughout the lattice and dramatically undo any hoped-for coordination. Chaotic systems are too chaotic to coordinate behavior between distant sites. The system cannot send a reliable signal across the lattice.

Conversely, suppose the lattice is deep in the ordered regime. A frozen red sea is spread across the lattice, leaving twinkling tiny green islands. Suppose we wish to coordinate a series of actions by distant sites. Alas, no signal can propagate across the frozen sea. The twinkling unfrozen islands are functionally isolated from one another. No complex coordination can occur.

But at the edge of chaos, the twinkling unfrozen islands are in tendrils of contact. Flipping any single lightbulb may send signals in small or large cascades of changes across the system to distant sites, so the behaviors in time and across the webbed network might become coordinated. Yet since the system is at the edge of chaos, but not actually chaotic, the system will not veer into uncoordinated twitchings. Perhaps, just perhaps, such systems might be able to coordinate the kinds of complex behavior we associate with life.

To complete this part of the story, I will present evidence for an idea that I will more fully develop in the next chapter: *the reason complex systems exist on, or in the ordered regime near, the edge of chaos is because evolution takes them there.* While autocatalytic networks arise spontaneously and naturally because of the laws of complexity, perhaps natural selection then tunes their parameters, tweaking the dials for K and P, until they are in the ordered regime near this edge—the transitional region between order and chaos where complex behavior thrives. After all, systems capable of complex behavior have a decided survival advantage, and thus natural selection finds its role as the molder and shaper of the spontaneous order for free. In order to test this hypothesis, Bill Macready, a postdoctoral fellow, Emily Dickinson, a computer scientist, and I have been using computer simulations to "evolve" Boolean networks to play simple and hard games with one another. In these games, each network must respond with a "proper" pattern of lightbulb activities to the prior pattern of lightbulb activities by the network it is playing. Our evolving networks are free to mutate connections between lightbulbs in each network and the Boolean rules turning lightbulbs on and off in each network. Thus our networks can change the different parameters that tune their positions on the order–chaos axis. In order to test the locations of our networks on the order–chaos axis, Bill, Emily, and I make use of a simple feature that distinguishes the ordered regime from the chaotic regime. In the chaotic regime, similar initial states tend to become progressively more dissimilar, and hence to *diverge* farther and farther apart in state space, as each passes along its trajectory. This is just the butterfly effect and sensitivity to initial conditions. Small perturbations amplify. Conversely, in the ordered regime, similar initial states tend to become more similar, hence *converging* closer together as they flow along their trajectories. This is just another expression of homeostasis. Perturbations to nearby states "damp out." We measure average convergence or divergence along the trajectories of a network to determine its location on the order–chaos axis. In fact, in this measure, networks at the phase transition have the property that nearby states neither diverge nor converge.

What are the results? As the networks play their games with one another, trying to match one another's lightbulb patterns, the computer simulation selects fitter mutant variants—that is, the networks that play better. What we have found for the modestly complex behaviors we are requesting is that the networks do adapt and improve and that they evolve, not to the very edge of chaos, but to the ordered regime, not too far from the edge of chaos. It is as though a position in the ordered regime near the transition to chaos affords the best mixture of stability and flexibility.

It is far too early to assess the working hypothesis that complex adaptive systems evolve to the edge of chaos. Should it prove true, it will be beautiful. But it will be equally wonderful if it proves true that complex adaptive systems evolve to a position somewhere in the ordered regime near the edge of chaos. Perhaps such a location on the axis, ordered and stable, but still flexible, will emerge as a kind of universal feature of complex adaptive systems in biology and beyond.

We turn to these beautiful possibilities in more detail in the following chapters, for the hypothesis that complex systems may evolve to the edge of chaos or to the ordered regime near that poised edge appears to account for a very large number of features of ontogeny, that magnificent, ordered dance of development from fertilized egg to bird, fern, bracken, flea, and tree. But caveats again, for at this stage a potential universal law is best held as a fascinating working hypothesis.

In the meantime, we may begin to suspect, the exquisite power of self-organization, which we begin to understand in our simple models of enormous Boolean networks, may be the ultimate wellspring of dynamical order. The order in these open nonequilibrium thermodynamic systems derives from the ordered regime; in turn, the order of the ordered regime derives from the fact that nearby states tend to converge. The system therefore "squeezes" itself onto tiny attractors. Ultimately, it is this self-squeezing into infinitesimal volumes of state space that constitutes the order. And while I have called it order for free, meaning that such order is natural and spontaneous, it is not "for free" thermodynamically. Rather, in these open systems, the self-squeezing of the system into tiny regions of state space is "paid for" thermodynamically by exporting heat to the environment. No laws of thermodynamics are violated or even contested. What is new is that vast open thermodynamic systems can spontaneously lie in the ordered regime. Such systems may be the natural source of the order required for stable self-reproduction, homeostasis, and graceful heritable variation.

If we, and past eons of scholars, have not begun to understand the power of self-organization as a source of order, neither did Darwin. The order that emerges in enormous, randomly assembled, interlinked networks of binary variables is almost certainly merely the harbinger of similar emergent order in whole varieties of complex systems. We may be finding new foundations for the order that graces the living world. If so, what a change in our view of life and our place must await us. Selection is not the sole source of order after all. Order vast, order ordained, order for free. We may be at home in the universe in ways we have hardly begun to comprehend.

Chapter 5

The Mystery
of Ontogeny

A t least since the Cambrian explosion, 550 million years ago, and probably for the past 700 million years, multicelled organisms have mastered a mystery no human mind yet comprehends: ontogeny. Through some mysterious evolutionary creativity, the new creatures of the Cambrian—and *Homo sapiens* much more recently—began life as a single cell, the zygote, the fruit of parental union. Somehow that single cell knew to give rise to a complete structure, an organized whole, an organism. If the swarm of stars in a spiral galaxy, clustered swirling in the high blackness of space, astonishes us with the wonder of the order generated by mutually gravitating masses, think with equal wonder at our own ontogeny. How in the world can a single cell, merely some tens of thousands of kinds of molecules locked in one another's embrace, know how to create the intricacies of a human infant? No one knows. If *Homo habilis* wondered, if Cro-Magnon wondered how they came to be, so too must we.

Begin, then, with the zygote. After fertilization of egg by sperm, the human zygote undergoes rapid cleavage—cell divisions that create a small mass of cells. These cells migrate down the fallopian tube and enter the uterus. While migrating, the mass of cells hollows out, forming a ball. A small number of cells, called the inner cell mass, migrates inward from one pole of the hollow ball and lodges nestled against the remaining outer layer. All mammals derive from the inner cell mass. The outer layer of cells in humans has specialized to burrow into the uterine lining and form the extraembryonic membranes, placenta and otherwise, that support us before birth.

Already, even at this most primitive stage, we witness the two fundamental processes of ontogeny, or development: the first is cell differentiation; the second is morphogenesis. The zygote is both a single cell and, necessarily, a single type of cell. Over the course of the 50 or so sequential rounds of cell division between the zygote and the newborn, that single cell gives rise to a menagerie of different cell types. The human body is thought to contain 256 different cell types, all specialized for specific functions in the tissues and organs. In broad terms, our tissues derive from three so-called germ layers: endoderm, mesoderm, and ectoderm. The endoderm yields cells and tissues of the intestinal tract, liver, and other tissues. A host of diverse cell types are formed, ranging from the specialized cells in the lining of the stomach, which secrete hydrochloric acid to help digest food, to the liver hepatocytes, which aid in detoxification of the blood. The mesoderm gives rise to muscle cells, the cells that form bone and cartilage, and the cells that form the blood, both the red blood cells carrying oxygen and the white cells of the immune system. The ectoderm layer gives rise to skin cells and the enormous diversity of nerve cells that form the peripheral and central nervous system.

In brief, the human zygote undergoes some 50 cell divisions, creating the 2^{50} or 10^{15} cells in your body. The initial zygote differentiates along branching pathways, ultimately giving rise to 256 diverse cell types that form the tissues and organs of the human infant. The increased diversity of cell types is called cell differentiation. Their coordination into organized tissues and organs is called morphogenesis.

I entered biology because the magnificent wonder of cell differentiation overwhelmed me. If I accomplished nothing else in this chapter except to convey this wonder, I would be content. I hope for more, however, because I believe that the spontaneous order discussed in previous chapters is the ultimate source of the order of ontogeny.

Recall the preformationist argument: the zygote contains a tiny homunculus that somehow expands to form the adult during development. Recall also the difficulty with any such theory, given a very large number of ancestors and a potentially infinite number of descendants. Recall that Hans Dreisch used a hair to separate the frog embryo at the two-celled stage and found that each cell gave rise to a perfect, if somewhat smaller, frog. How could both cells retain the information to generate the entire frog?

The frog is not alone in its mastery of this trick. The carrot is even fancier in its capacity to develop. If the carrot is broken into single cells, then virtually any one of them, whatever its type, can regenerate the whole plant. How can all the cells become different, yet each retain the information needed to form the entire organism?

Earlier in this century, soon after the rediscovery of Mendel's laws and the establishment of the theory that chromosomes carry the genes, it had been supposed that the zygote had the full complement of genes, but that they were parsed out in different subsets to different cell types in the organism. Only the germ plasm, giving rise to sperm or egg, would retain the full complement of genes. But it soon became clear, based on microscopic studies of cells, that with rare exceptions, all the cell types of an organism contain the full complement of chromosomes. All the cells carry all the genetic information that the zygote has. More recent work, at the level of DNA, supports this broad claim. Almost all cells of almost all multicelled organisms contain the same DNA, but there are exceptions. In some organisms, the entire paternal chromosome set is lost. In some cells in some organisms, certain genes are replicated several extra times. In immune-system cells, the chromosomes are rearranged and modified slightly to generate all the antibodies needed to fight off intruders. But, by and large, the sets of genes in all the cells of your body are identical.

The growing awareness that all the cells in a multicelled organism have the same set of genes has led to what might be called the central dogma of developmental biology: cells differ because different genes are active in the different cell types of the organism. Thus red blood cells express the gene that encodes hemoglobin. B-cells in the immune system express the genes that encode antibody molecules. Skeletal-muscle cells express the genes that encode actin and myosin molecules, which form muscle fibers. Nerve cells express the genes for the proteins that form specific ion-conducting channels in the cell membrane. Certain gut cells express the genes that encode enzymes leading to synthesis and secretion of hydrochloric acid.

But what is the mechanism that allows some genes to be active while others are suppressed? And how, as the zygote unfolds into the body, do the various cell types know which proteins to express?

Jacob, Monod, and Genetic Circuits

Two French biologists, François Jacob and Jacques Monod, won the Nobel Prize in the mid-1960s for the work that has provided us with the start of a conceptual framework to account for cell differentiation and ontogeny.

As noted earlier, the synthesis of a protein requires that the gene encoding it be transcribed from DNA to RNA. The corresponding messenger RNA is then translated via the genetic code into a protein. Jacob and Monod made their discovery by studying the behavior of the gut bacterium, *E. coli,* and its response to a sugar called lactose. It was known that when lactose is first added to the medium, the cells are unable to utilize the molecule. The appropriate enzyme for lactose breakdown, beta-galactosidase, is not present in the bacterial cell at sufficient concentration. Within a few minutes of the addition of lactose, however, *E. coli* cells begin to synthesize beta-galactosidase and then begin to use lactose as a carbon source for cell growth and division.

Jacob and Monod soon discovered how such enzyme induction, the term used for the capacity of lactose to induce the synthesis of beta-galactosidase, is controlled. It turned out that control occurs at the level of transcription of the beta-galactosidase gene into the corresponding messenger RNA. Adjacent to this structural gene—so named because the gene encodes the structure of a protein—Jacob and Monod discovered that there is a short sequence of nucleotides in the DNA to which a protein binds. The short sequence is called the operator. The protein binding to the operator is called the repressor. As the name implies, when the repressor protein is bound to the operator site, transcription of the beta-galactosidase gene is repressed. Hence no messenger RNA is formed for this enzyme, and the enzyme itself is not formed by translation of the messenger RNA.

Now comes the regulatory magic. When lactose enters the *E. coli* cell, it binds to the repressor, changing its shape so that the repressor can no longer bind to the operator. Thus the addition of lactose frees the operator site. Once the operator is free, transcription of the adjacent beta-galactosidase structural gene can begin, and the beta-galactosidase enzyme is soon produced.

Jacob and Monod had discovered that a small molecule can "turn a gene on." Because the repressor itself is the product of another *E. coli* gene, it soon became obvious that genes can form genetic circuits and turn one another on and off. By 1963, Jacob and Monod had written a groundbreaking article suggesting that cell differentiation would be controlled by just such genetic circuits. In the simplest case, two genes might each repress the other. Think a moment about such a system. If gene 1 represses gene 2, and gene 2 represses gene 1, then such a system might have two different patterns of gene activity. In the first pattern, gene 1 would be active and repress gene 2; in the second, gene 2 would be active and repress gene 1. With two different stable patterns of gene expression, this little genetic circuit would be able to create two different cell types. Each of these cells would be an alternative pattern of the same genetic circuit. Both cell types would then have the same "genotype," the same genome, but be able to express different sets of genes.

Jacob and Monod had broken a hammerlock. Not only did their work begin to suggest how cell differentiation might occur, but it revealed an unexpected and powerful molecular freedom. The repressor protein binds to the operator site using a specific site on the repressor protein. The lactose molecule (actually a metabolic derivative of lactose called allolactose) binds to a second site on the repressor protein. The binding of the allolactose molecule changes the shape of the repressor protein, thereby changing the shape of its first site, thereby lowering the affinity of the repressor protein for the operator DNA site. So allolactose, binding a second site on the repressor, "pulls" it off the operator, thereby allowing synthesis of the gene that metabolizes lactose, beta-galactosidase. But since allolactose acts via a second site, called an allosteric site, on the repressor protein, which is different from the site on the repressor that binds to the operator DNA sequences, this implies that the shape of the allolactose molecule need bear no obvious relation to the ultimate consequences of its action—its capacity to control gene activity. In contrast, a substrate must fit its enzyme, and a second molecule, a competitive inhibitor that is to inhibit the enzyme by binding to the same enzymatic site, must look like the real substrate. Here, in this familiar case, the substrate and competitive inhibitor necessarily have similar molecular features. But because allolactose acts at a second site on the repressor rather than at the repressor's DNA binding site, allolactose might just as well be used as a signal to control the transcription of a gene encoding actin, myosin, or an enzyme involved in the synthesis of hydrochloric acid. The shape of the molecular-control molecule need bear no relation to the ultimate product of the controlled process. As both authors stressed, action via second sites meant utter freedom from the molecular point of view to create genetic circuits of arbitrary logic and complexity.

Selection as the Sole Source of Order

So struck was Monod by the freedom to construct arbitrary genetic circuits that he wrote a beautiful book, *Chance and Necessity*. In it Monod carved the sweet, lyric phrase I have mentioned before: "Evolution is chance caught on the wing." This phrase, more than any I know, captures our sense, since Darwin, of untold freedom for the search of random mutation, of selection culling the rare useful forms from the chaff of useless ones.

Since Darwin, as I have already stressed, we have come to view selection as the only source of order in organisms. No tiny matter, for this image lies at the heart of our sense of all life, all organisms, all humans, as profoundly contingent, as historical accidents. Jacob noted, as men-

tioned previously, that evolution is an opportunist that creates bricolage, tinkered-together contraptions. We, the ad hoc solutions to histor-
ically mandated design problems. We, the molecular Rube Goldbergs
of this era, descendants of Rube Goldbergs before us.

But if selection, working on random variations, is the sole source of
order, then we stand twofold stunned: stunned because the order is so
very magnificent; stunned because the order must be so very unexpected, so rare, so precious. We the unexpected, orphaned in the spell-
binding vastness of space. But has selection truly acted alone as the sole
source of order in the emergence of life and its subsequent evolution? I
do not think so. From my gut, from my dreams, from my work of three
decades, from the work of a growing number of other scientists, I do
not think so.

I came to biology from philosophy, psychology, and physiology. I had
been fortunate enough to graduate from Dartmouth College six months
before formal graduation in June 1961. After six months as the requisite
mountain-climbing, ski-crew member grooming the slopes, living (in a
Volkswagen camper I crudely outfitted) at the finest address in St.
Anton, Austria—the parking lot of the Post Hotel—I entered Oxford
on a Marshall Scholarship.

My teachers, philosopher Geoffrey Warnock and psychologist Stuart
Sutherland, prized on-the-spot invention. Was language prior to recognition? How could a neural circuit allow the eye to distinguish the off-
set of two parallel lines when the distance between them was less than
the width of a single cone or rod in the retina? One got training at invention, a training allowed and supported by an English tradition worth
emulating: the British love their eccentrics. One don used to sing Gre-
gorian chants in the bathroom. Such an environment nurtures a spirit
that is also admired by many physicists. Upon hearing a theory by a
younger colleague, Wolfgang Pauli is said to have responded, "It's crazy,
but not crazy enough!" If Monod and Jacob saw unbridled molecular
freedom at the foundations of evolution, then ought we not seek unbridled intellectual freedom at the foundations of the scientific enterprise?
Always allow ourselves and our colleagues to be crazy enough. The
world will tell us if we are right or wrong.

At its deepest roots, science grows by the questions we ask. What are
the sources of those questions? I do not know. But I do know that I
have always hoped that the order in organisms would come to be understood as natural, as expected. I have always harbored the dream that
selection has always had a partner in molding life: self-organization.

If selection had to search too hard to find a genetic mechanism that
allowed the zygote to unfold so beautifully, then the result would be too

ad hoc, one of Jacob's tinkered-together contraptions. Thus as a medical student, I found myself hoping that large networks of genes would spontaneously exhibit the order necessary for ontogeny. That there was a sacredness, a law—something natural and inevitable about us.

The Spontaneous Order of Ontogeny

And, as shown in Chapter 4, the self-organization of which I dreamed abounds. I suggest that order, order for free, is the ultimate wellspring of the order of ontogeny. I must warn you that this is a heretical view. But the spontaneous order we have already encountered is so powerful that it seems simply foolish or stubborn not to examine with the utmost seriousness the possibility that much of the order of ontogeny is spontaneous, crafted thereafter by selection.

Jacob and Monod had told a waiting community of biologists that genes can turn one another on and off, that genetic circuits can have alternative patterns of gene activities that constitute the different cell types of an organism. What is the structure of such genetic networks? What are the rules governing the behaviors of the genes and their products coupled together in the webs of control that govern ontogeny?

I invented the Boolean network models discussed in Chapter 4 specifically to explore these questions. Recall that we were concerned with lightbulbs that turn one another on and off, and I interpreted these lightbulbs as enzymes promoting or inhibiting one another's production. But the same ideas apply to genetic regulatory circuits of the Jacob and Monod type. We can think of the structural gene for beta-galactosidase as either on or off, either being or not being transcribed; we can think of the repressor protein as either bound or not bound to the operator site, as on or off. We can think of the operator site as free or not free, as on or off. We can think of allolactose as having bound or not having bound the second site on the repressor protein. While this is surely an idealization, we can extend it to networks of genes and their products interacting with one another in enormous webs of regulatory circuitry.

We can, in sum, model a genetic regulatory system as a Boolean network. The "wiring diagram" among the lightbulbs now means the molecular regulatory connections among the genes and their products. In the current context, the repressor protein is a molecular regulatory input to the operator, while the operator itself is a regulatory input to the activity of the beta-galactosidase gene. The Boolean functions, or rules, showing which patterns of lightbulbs turn a regulated lightbulb on or off, now mean the combinations of molecular signals that activate or inhibit a given gene activity. For example, the operator is controlled by both the repressor and allolactose (Figure 5.1). The operator is bound by the repressor unless allolactose binds to the repressor and pulls the repressor off the operator. So the operator is controlled by the Boolean NOT IF function. The gene producing beta-galactosidase is inactive, but *not if* allolactose is present.

In Chapter 4, it was shown that Boolean network models, or genetic network models as I will now call them, with N genes can be in one of 2^N states, in which a different combination of genes is turned on and off. In the language of dynamical systems, its state space consists of 2^N different possible gene activities. Recall that as the genes follow the Boolean rules and switch on and off according to the activities of their molecular inputs, such a network follows a trajectory in its state space. Ultimately, the trajectory converges onto a state-cycle attractor around which the system will cycle persistently thereafter. A variety of different trajectories may all converge on the same state cycle, like water draining into a lake. The state-cycle attractor is the lake, and the trajectories converging onto it constitute its basin of attraction. Any Boolean network, and therefore any model genomic regulatory network, must have at least one such state-cycle attractor, but may have many, each draining its own basin of attraction.

The purpose of Chapter 4—indeed, the hope of my early research into network models—was to see whether vast networks with thousands of coupled lightbulbs might spontaneously exhibit order. We found massive order. We found that networks with as many as 100,000 bulbs, and state spaces of $2^{100,000}$, hence $10^{30,000}$, would settle down and cycle through a tiny, tiny state cycle with a mere 317 states on it. Order for free. As I said, 317 compared with $10^{30,000}$ corresponds to the network squeezing itself into a fraction of its state space equal to 1 divided by $10^{29,998}$.

I cannot show you an attractor in such an unfathomable state space. Figure 5.2 shows four attractors in a small network with only 15 lightbulbs that can be in about 32,000 states or patterns of illumination. Imagine each attractor as a pointlike black hole in the vastness of network state space, sucking in everything that falls into its basin of attraction. The entire unthinkably large state space breaks up into some modest number of these black holes, each commanding the entire flow from megaparsecs of the state space around it. Release the system anywhere and, like a spaceship hurtling toward its final omega point, the system plunges swiftly toward the tiny point of space, the attractor, which draws it ineluctably forward.

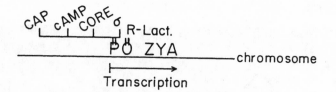

allo-lactose	Repressor	Operator
0	0	0
0	1	1
1	0	0
1	1	0

"NOT IF"

CAP	cAMP	CORE	SIGMA	PROMOTER
0	0	0	0	0
0	0	0	1	0
0	0	1	0	0
0	0	1	1	0
0	1	0	0	0
0	1	0	1	0
0	1	1	0	0
0	1	1	1	0
1	0	0	0	0
1	0	0	1	0
1	0	1	0	0
1	0	1	1	0
1	1	0	0	0
1	1	0	1	0
1	1	1	0	0
1	1	1	1	1

"AND"

Figure 5.1 *Genetic circuitry. The upper panel shows the lactose operon in the bacterium* E. coli. *Z, Y, and A are structural genes; O is the operator site. P is the promoter site, and R is the repressor protein that binds to the operator and blocks transcription unless it is itself bound by lactose or its metabolite, allolactose. (The promoter is regulated by four molecular factors: cyclic AMP, core RNA polymerase, sigma factor, and CAP.) The middle panel shows a Boolean function describing regulation of the operator by repressor and allolactose. For the operator site, 0 = free, 1 = bound. For repressor and lactose, 0 = absent, 1 = present. The Boolean function* NOT IF *specifies the activity of the operator at the next moment, given each of the four possible current states of the regulatory inputs. The lower panel shows the regulation of the promoter by its four molecular inputs. The Boolean function is the* AND *function of four variables.*

Figure 5.2 *Basins of attraction. Four basins of attraction and attractor state cycles in the repertoire, or behavior field, of one random Boolean network with* N = 15 *binary variables, and* K = 2 *inputs per binary variable.*

Tiny attractors. Vast, vast order.

Jacob and Monod suggested that the alternative stable patterns of their genetic circuit—gene 1 on, gene 2 off and gene 1 off, gene 2 on—were the different cell types of one genomic network. I suggest that each black-hole state-cycle attractor in the vast state space of a genomic network is a different cell type. A network consisting of even a modest number of genes can potentially explore a vast state space. But, if I am correct, a handful of attractors pull it in a few directions. Depending on which state cycle the network is orbiting, different genes will be switched on and off. Different proteins will be produced. The genomic network will act as a different kind of cell.

A great many predictions follow from this single hypothesis. A great many features of genomic systems and ontogeny seem to fall into place, each comfortable in this new conceptual framework. So while the new hypothesis remains to be proved, a lot of evidence already supports it.

Let us restate the problem of ontogeny. The human genome encodes some 100,000 structural genes and an unknown number of operators, repressors, promoters (which switch other genes on instead of off), and so forth. These genes, together with their RNA and protein products, are connected into a tangled web of regulatory interactions, the genomic regulatory network, whose joint behavior coordinates development from the zygote to the adult. The human body has 256 cell types. A fruit fly, *Drosophila melanogaster,* has something like 15,000 structural genes and perhaps 60 cell types. Hydra has a very much smaller genome and some 13 to 15 cell types. What principles marshal such genetic regulatory networks to the exquisite order of ontogeny?

If the genetic circuits of bacteria and higher organisms are examined, three features emerge:

1. Any gene or other molecular variable is directly regulated by rather few molecular inputs. For example, the lactose operator described earlier is regulated by two molecular inputs: allolactose and the repressor protein.

2. The Boolean rules describing the activities of different genes differ. Thus the lactose operator is derepressed by lactose according to the NOT IF Boolean rule. Other genes are activated by molecular inputs according to the OR or the AND Boolean rules, or follow more complex rules.

3. Using the idealization that genes are binary variables, active or inactive in transcription, while their control inputs are present or absent, hence using the idealization of Boolean functions introduced in Chapter 4, known genes are regulated by a special subset of Boolean functions that I will call canalyzing functions and will characterize more precisely in a moment.

Here is a surprising fact: almost any genomic regulatory network harboring these three known properties will exhibit all the order for free we might hope for. These known properties already predict much of the order of the biological world.

In Chapter 4, we saw that large networks of binary elements, random Boolean networks, generically behave in three regimes: a chaotic regime, an ordered regime, and the edge-of-chaos complex regime. We saw that two simple constraints suffice to guarantee that most members of the constrained ensembles lie in the ordered regime. It suffices if each binary element receives $K = 2$ or fewer inputs. Or, if the network has more than $K = 2$ inputs per lightbulb, then certain biases in the Boolean rules, captured by the P parameter, can be adjusted to ensure order.

Another way to ensure orderly behavior is to construct networks using what are called canalyzing Boolean functions. These Boolean rules have the easy property that at least one of the molecular inputs has one value, which might be 1 or 0, which by itself can completely determine the response of the regulated gene. The OR function is an example of a canalyzing function (Figure 5.3*a*). An element regulated by this function is active at the next moment if its first, or its second, or both inputs are active at the current moment. Thus if the first input is active, then the regulated element is guaranteed to be active at the next moment, regardless of the activity of the second input. This property defines a canalyzing Boolean function. At least one input must have one value, 1 or 0, which alone suffices to guarantee that the regulated variable will have one value.

Almost all regulated genes in viruses, bacteria, and higher organisms of which I am aware are governed, in the Boolean idealization, by canalyzing Boolean functions. As shown in Figure 5.1, the operator site is governed by the canalyzing NOT IF Boolean function. If the repressor is absent, then the operator site is guaranteed to be free, regardless of the presence or absence of allolactose. If allolactose is present, the operator is also guaranteed to be free regardless of the presence or absence of the repressor protein. This is true because allolactose binds the allosteric site on the repressor, pulling it off the operator site.

Most Boolean functions with many inputs are not canalyzing. That is, they do not have the property that any single input can, by itself, determine the next state of the regulated lightbulb. The simplest example is the EXCLUSIVE OR function with two inputs (Figure 5.3*b*). A gene regulated by this function will be active at the next moment if either one, but not both, of the inputs is active at the current moment. As you can see, no activity of either input, 1 or 0, suffices by itself to guarantee the activity of the regulated gene. Thus if the first input is 1, the regulated gene can be active if the second input is 0, or can be inactive if the second input is 1. If the first input is 0, the regulated gene can be active if the second input is 0, or can be inactive if the second input is 1. The

A	B	C
0	0	0
0	1	1
1	0	1
1	1	1

a

A	B	C
0	0	0
0	1	1
1	0	1
1	1	0

b

Figure 5.3 *Boolean functions.* (a) *The Boolean* OR *function of two inputs. If A or B (or both) is 1, then C is 1.* (b) *The Boolean* EXCLUSIVE OR *function of two inputs. If A or B (but not both) is 1, then C is 1.*

same story holds for the second input. It has no activity, which guarantees the behavior of the regulated gene.

It is probably not an accident that regulated genes, and most other biochemical processes, appear to be governed by canalyzing functions, for canalyzing functions are rare among the possible Boolean functions, becoming rarer as the number of inputs, K, increases. But they are simple to build chemically. Thus the abundance of canalyzing functions reflects either selection for a rare kind of Boolean rule or chemical simplicity. Either way, the abundance of canalyzing functions appears to be very important to the ordered behavior of genomic regulatory systems.

The number of possible Boolean functions with K different inputs is $2^{(2^K)}$. This is simple to see. With K inputs, there are 2^K possible combinations of activities. A Boolean function must choose a 1 or a 0 response for each of these input combinations, hence the formula given. The fraction of these functions that are canalyzing is very high for $K = 2$; indeed, 14 of the 16 Boolean functions of $K = 2$ inputs are canalyzing (Figure 5.4). Only two, EXCLUSIVE OR and its complement, a function

1	2	3
0	0	0
0	1	0
1	0	0
1	1	0

1	2	3		1	2	3		1	2	3		1	2	3
0	0	0		0	0	0		0	0	0		0	0	1
0	1	0		0	1	0		0	1	1		0	1	0
1	0	0		1	0	1		1	0	0		1	0	0
1	1	1		1	1	0		1	1	0		1	1	0

1	2	3		1	2	3		1	2	3		1	2	3		1	2	3		1	2	3
0	0	0		0	0	0		0	0	1		0	0	1		0	0	1		0	0	0
0	1	0		0	1	1		0	1	0		0	1	0		0	1	1		0	1	1
1	0	1		1	0	0		1	0	0		1	0	1		1	0	0		1	0	1
1	1	1		1	1	1		1	1	1		1	1	0		1	1	0		1	1	0

1	2	3		1	2	3		1	2	3		1	2	3
0	0	1		0	0	1		0	0	1		0	0	0
0	1	1		0	1	1		0	1	0		0	1	1
1	0	1		1	0	0		1	0	1		1	0	1
1	1	0		1	1	1		1	1	1		1	1	1

1	2	3
0	0	1
0	1	1
1	0	1
1	1	1

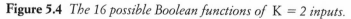

Figure 5.4 *The 16 possible Boolean functions of* $K = 2$ *inputs.*

called IF AND ONLY IF, are not canalyzing. But among the 64,000 or so Boolean functions of $K = 4$ inputs, only 5 percent are canalyzing. As K increases, the fraction that are canalyzing falls still further.

Canalyzing functions are simple to build from a molecular point of view. Consider an enzyme with two inputs, activated if either one is active. This is easy to construct; we need make an enzyme with only a single allosteric site. The binding of either molecular input to that site alters enzyme conformation and activates the enzyme. This is the canalyzing OR function. But how might one make an enzyme that realizes the noncanalyzing EXCLUSIVE OR function? This would require two distinct allosteric sites. The binding of either site alone by its molecular input, or effector, would have to alter the enzyme to allow activity. But the binding of both allosteric sites simultaneously, or of neither, could not allow the same change that activates the enzyme. Such molecular machinery is surely possible, but surely more difficult to achieve than the OR function. In general, it seems easier to create molecular devices that realize the canalyzing functions rather than the noncanalyzing functions.

The importance of the apparent chemical simplicity of canalyzing functions is this: large networks of binary elements governed predominantly by canalyzing functions spontaneously lie in the ordered regime. Vast order for free abounds for selection's further sifting. If canalyzing functions are abundant in cells because they are chemically simple, then that chemical simplicity itself suffices to engender massive large-scale spontaneous order.

I believe that this spontaneous order is crucial to understanding the behavior of the genome. Since each of our cells houses some 100,000 or more genes, the state space of the human genomic regulatory system is at least $2^{100,000}$ or $10^{30,000}$. As we have noted, this number is meaninglessly enormous compared with anything we know about. In terms of this vast state space, what is a cell type? The central dogma of developmental biology merely states that different cell types are different patterns of activity of the same genomic system. That is not much help when the human genome affords at least $10^{30,000}$ combinations of gene activity. The range of possibilities is even greater if we drop the on–off idealization and remember that genes are capable of graded levels of expression, enzymes of graded levels of activity. With either on–off idealization or graded levels of activity, cells could not possibly explore such a range of patterns of gene activities in the lifetime of all the creatures that might exist on all the worlds that might exist since the Big Bang.

This mystery begins to yield if we entertain the possibility that, because of the way they are constructed, genomic networks lie in the ordered regime. Instead of roaming all over the map, such a network is pulled by a handful of attractors, black holes in the genome's state space. A cell orbiting a particular attractor will express certain genes and proteins, making it behave as a certain type of cell. The same cell orbiting a different attractor will express other genes and proteins. Thus our framework hypothesis states that cell types are attractors in the repertoire of the genomic network.

Within this framework, many known properties of ontogeny fall readily into place. First of all, each cell type must be confined to an infinitesimal fraction of the possible patterns of gene activity. Just this behavior arises spontaneously in the ordered regime. The lengths of state-cycle attractors scale as the square root of the number of genes; thus a human genomic system with 100,000 genes and $10^{30,000}$ possible patterns of gene expression should settle into and cycle around state cycles with only about 317 states, an infinitesimal fraction of the possible patterns of gene activity. The small attractors of the ordered regime constitute order for free.

The predicted time it should take cells to orbit their attractors is completely biologically plausible. It takes on the order of one to 10 minutes to turn a gene on or off. Thus the time to orbit the state cycle should be 317 minutes to 3,170 minutes, or from about five to 50 hours—precisely in the plausible range for cell behavior!

In fact, the most obvious cycle that cells undergo is the cell-division cycle itself. In bacteria at full speed, the cell cycle takes about 20 minutes. There are cells lining the intestine in an area called the crypts of Leberkuhn that cycle every eight hours; other cells in your body cycle in something like 50 hours. So if cell types correspond to state-cycle attractors, then a cell cycle can be thought of as a cell traversing the length of its state cycle. And the resulting time scale to traverse the attractor is the real time scale of the cell cycle.

A genetic network with two inputs per gene ($K = 2$), or a network rich in canalyzing functions, exhibits not only spontaneous order, but order that is similar to that found in real cells. Remember from Chapter 4 that in a $K = N$ network with 100,000 genes, a cycle would have $10^{15,000}$ states on it. By now you can calculate how long it would take to complete an orbit around a state-cycle attractor at a minute per state transition. From the biological perspective, forget it. Even networks with $K = 4$ or $K = 5$, which are already well into the chaotic regime, have exorbitantly long state cycles. Here we consider genomic networks built entirely at random, subject only to known constraints found in real genomic networks, and we find cycle times exactly in the biological ballpark.

If this view is right, then cell-cycle times should scale as about a square-root function of the number of genes. Figure 5.5 shows this to be true for organisms ranging from bacteria to yeast to hydra to humans. That is, the median cell-cycle time varies as about a square-root function of the number of genes in that organism. So for a bacterium, the dividing time is about 20 minutes, while for a human cell, with 1,000-fold more DNA per cell, dividing time is about 22 to 24 hours.

Figure 5.5 allows us to see that median cell-cycle times do actually increase as about a square root of the number of genes in an organism, as predicted by our hypothesis. But Figure 5.5 also allows us to say—as our model also predicts—that the distribution around the median is strongly skewed, with most cells at any genomic complexity having short cycle times, while a few have long cycle times. The same skewed distribution is found in model genomic networks with $K = 2$ inputs per gene. But I owe you a cautionary note: whereas a great deal of work has been done on the cell cycle, we do not know enough to say more than

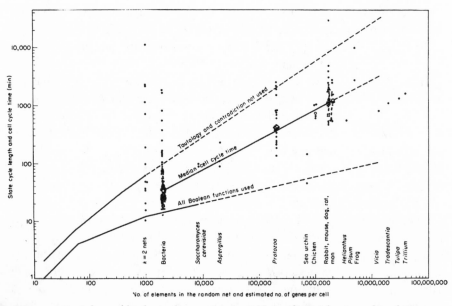

Figure 5.5 *A law of biological organization? Using logarithmic scales, the number of elements in a random network is plotted against the length of the state cycle, and the estimated number of genes for various organisms, assumed proportional to DNA content per cell, is plotted against the cells' median replication times. In both cases, the result is a straight line with a slope of 0.5. This is the hallmark of a square-root relation. (Tautology and Contradiction, the first and last Boolean functions shown in Figure 5.4, were not used in the networks.)*

that a powerful statistical correspondence holds between theory and observation.

We will now see more of these powerful correspondences. The number of cell types in organisms increases from one or two in bacteria, to three in yeast, to perhaps 13 to 15 in a simple organism like hydra, to perhaps 60 in the fruit fly, to 256 in you and me. At the same time, the number of genes increases from bacteria to humans. It would be wonderful to understand why different genomic systems of different complexities have the number of cell types they have.

If a cell type is a state-cycle attractor, then we should be able to predict the number of cell types as a function of the number of genes in an organism. With $K = 2$ inputs per gene, and more generally with canalyzing networks, the median number of state-cycle attractors is only about the square root of the number of genes. Thus our prediction: a human, with 100,000 genes, should have about 317 cell types. And, in fact, the number of known human cell types is 256.

If our theory is right, we should be able to predict the scaling relation between the number of genes and the number of cell types. The latter should increase as a square-root function of the former. Figure 5.6, with

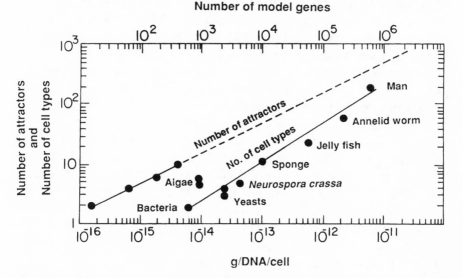

Figure 5.6 *Another candidate law. The logarithm of the number of cell types in organisms across many phyla is plotted against the logarithm of the DNA content per cell. Again, the plot is linear, with a slope of 0.5, indicating that the number of cell types increases as the square root of the amount of DNA per cell. If the number of structural and regulatory genes is assumed proportional to the DNA content per cell, the number of cell types increases as a square-root function of the number of genes.*

numbers of genes on the x-axis and numbers of cell types on the y-axis, confirms this prediction. The number of cell types does indeed increase about as a square-root function of the number of genes.

It is hard not to be impressed. The theory that genomic systems should lie in the ordered regime is once again not only in the biological ballpark, but almost exactly on the mark. Why should a theory of the generic behaviors of parallel-processing genomic networks in the ordered regime manage to predict the scaling relationship of cell types to genomic complexity, even roughly, and why should the absolute values be so close to that observed?

As noted, homeostasis, the tendency of cell types to remain the same following perturbations, is essential to life. Take a Boolean network with thousands of variables, and allow it to settle to a state-cycle attractor. Transiently "flip" the activity of any single model gene. For almost all such perturbations, the system returns to the state cycle from which it was perturbed. This is precisely homeostasis, and it comes for free in the ordered regime.

But homeostasis cannot be complete. If the zygote differentiates through branching pathways to intermediate cell types that themselves branch to the final cell types of the newborn or the adult, then occasionally a perturbation will have to push a cell into a new basin of attraction flowing to a new attractor—that is, into a new developmental pathway flowing to a new cell type. For example, in the early embryo, cells of the ectodermal germ layer that are on the pathway to form skin cells are known to be induced by a molecular trigger that causes those cells to jump to a new pathway and form nerve cells. At the same time, we know from nearly a century's work that each cell type can be triggered to change into only a few neighboring pathways. Ectodermal cells in young embryos can switch from skin to nerve, but not to the cells lining the stomach and secreting hydrochloric acid.

Hold on to the picture of the zygote dividing and sending daughter cells down branching pathways, with two or three options at each branch point, to create the ultimate diversity of cell types. As far as we know, all multicelled organisms have always developed via such branching pathways of differentiation. Do our model genomic systems in the ordered regime naturally exhibit these properties? Yes, they do.

For most perturbations, a genomic system on any attractor will exhibit homeostatic return to the same attractor. The cell types are fundamentally stable. But for some perturbations, the system flows to a different attractor. So differentiation occurs naturally. And the further critical property is this: from any one attractor, it is possible to undergo transitions to only a few neighboring attractors, and from them other perturbations drive the system to still other attractors. Each lake, as it were, is close to only a few other lakes. An ectodermal cell can be easily bumped into an attractor where it will form a retinal cell, but the ectodermal cell cannot be easily bumped into the basin where it would form a gut cell.

Without further selection, genomic networks in the ordered regime spontaneously have a fundamental property that has characterized ontogeny for perhaps a billion years: cells differentiate down branching pathways from the zygote to yield the many cell types of the adult. Has selection struggled for a billion years to maintain this core feature of ontogeny? Is this feature shared among all multicelled organisms by virtue of common descent? Or are branching pathways of differentiation a feature so deeply embedded in canalyzing genomic networks that this profound aspect of ontogeny shines through as an expression of order for free, whatever selection's further sifting? If so, then selection is not the only source of order in ontogeny.

There is more evidence as well. Recall that in the ordered regime, a "red" component of genes frozen in either the active or the inactive state forms a giant cluster spreading across the genomic network, leaving behind functionally isolated "green" islands of genes twinkling on and off in complex patterns. If the genomic system is indeed in the ordered regime, then such a frozen component and isolated unfrozen islands should occur. If so, then a large fraction of genes should be in the same state of activity in all cell types in the body. These should correspond to the "frozen component" spanning across the network. In fact, about 70 percent of the genes are thought to be a common core, simultaneously active in all cell types of a mammalian organism. Similar numbers arise in plants. These may be parts of a frozen component.

The further implication of this is that only a fraction of the genes determine the differences between cells. In a plant with something like 20,000 genes, the typical differences in gene expression between different cell types is found to be on the order of 1,000 genes, or 5 percent. This is very close to the fraction predicted from the expected number of genes in the unfrozen twinkling islands.

Finally, recall that in the ordered regime, perturbing the activity of a single gene propagates to only a small fraction of the genes in the entire network. Further, virtually no gene should unleash an avalanche of alterations that cascades to tens of thousands of other genes. This also is true.

Dreams of a Final Theory

In 1964, when entering medical school, I found myself with a dream whose origins I do not know. I was new to biology then, as unfamiliar as many of you may be with its intertwined marvels of historical contingency, selection, design, drift, accident, and sheer wonder. I think as a young scientist I could not yet begin to fathom the power of natural selection, whose subtlety has grown more impressive to me over the intervening three decades. Yet the dream that welled up from whatever unknowable sources is one I still hold. If biologists have ignored self-organization, it is not because self-ordering is not pervasive and profound. It is because we biologists have yet to understand how to think about systems governed simultaneously by two sources of order. Yet who seeing the snowflake, who seeing simple lipid molecules cast adrift in water forming themselves into cell-like hollow lipid vesicles, who seeing the potential for the crystallization of life in swarms of reacting molecules, who seeing the stunning order for free in networks linking tens upon tens of thousands of variables, can fail to entertain a central thought: if ever we are to attain a final theory in biology, we will surely, surely have to understand the commingling of self-organization and selection. We will have to see that we are the natural expressions of a deeper order. Ultimately, we will discover in our creation myth that we are expected after all.

Chapter 6

Noah's Vessel

A ncient Noah, forewarned of the coming flood, measuring in cubits and hewing good wood, built an ark for all species. Two by two, male and female, they filed solemnly up the sturdy gangplank and into the ark to await the coming of the waters. Crowded with the munificence of God's creation, we're told, the ark floated to touch on the flank of Mount Ararat and loose again the wonder of his works upon the lands.

Had there really been a Noah to take a census of the earlier biosphere, and if we could somehow gain access to what he found, we would presumably find that the diversity of molecules and the diversity of organisms have greatly increased since the true birth of our planet 4 billion years ago. The diversity of organic molecules is presumed to have been very low when the protoplanet formed. The diversity of species shortly after life's inception was presumably very low.

Countless kinds of organic molecules now swarm in the millions of kinds of cells that lie within a square mile of most places people dwell. No one knows how great that diversity of inorganic and organic molecules is with much precision, but we can be very sure of this: the diversity of organic molecules in the biosphere is vastly greater than it was 4 billion years ago when the first small molecules assembled themselves in the primal atmosphere and seas. Somehow, in some as yet mysterious process, the organic molecular diversity of this spinning globe has taken energy—in the forms of snippets of sunlight caught on the fly, hydrothermal energy sources, or lightning bolts—and cooked itself up from simple atoms and molecules to the complex organic molecules we find today.

In our search for hints that the order of the biosphere is shaped by laws deeper than natural selection alone, we now seek to understand

113

the wellsprings of this stunning molecular diversity. The striking possibility is that the very diversity of molecules in the biosphere causes its own explosion! The diversity feeds on itself, driving itself forward. Cells interacting with one another and with the environment create new kinds of molecules that beget yet other kinds of molecules in a rush of creativity. This rush, which I will call supracritical behavior, has its source in the same kind of phase transition to connected webs of catalyzed reactions that we found may have pushed molecules into living organizations in the first place.

In a nuclear chain reaction, decay of a uranium nucleus creates several neutrons. Each neutron may collide with another uranium nucleus to create a shower of still more neutrons, which in turn bombard yet more nuclei. The chain reaction feeds on itself, neutrons generating more neutrons generating more neutrons until an ominous cloud mushrooms skyward. In supracritical chemical systems, molecular species beget molecular species beget molecular species until the wondrous cloud coalesces into everything from trilobites to flamingos.

We are hunting big game, seeking laws of complexity, laws that govern the creative processes in this nonequilibrium expanding universe, with its abundance of energy coiled to form galaxies, complex molecules, and life. We have seen hints already. In Chapter 3, we explored the possibility that sufficiently diverse mixtures of chemicals reacting with one another can "catch fire," achieve catalytic closure, and suddenly emerge as living, self-reproducing, evolving metabolisms. Autocatalytic sets can crystallize as part of this order for free. In Chapters 4 and 5, we have seen further footprints of order for free in the startling and coherent dynamical order in lightbulb networks, and hence in autocatalytic networks of molecules and in today's cells and their ontogeny. The order of tiny attractors steers such molecular systems to their own coherence. But neither protocells nor today's cells live alone. Cells live in complex communities that always have traded and always will trade the molecules each cell creates. The ecosystem outside our windows, familiarly seen in terms of the kinds of species that form it, is simultaneously a network of metabolisms creating and trading their stuff. The ecosystems of the planet link into what may be the most complex chemical production factory that exists in our neck of the cosmos—the machine by which the nonequilibrium processes of our hunk of the universe swell the diversity of molecular forms, ensuring that complexity and creativity lie everywhere.

If today we could collectively assemble all the beasts of the land and fishes of the waters, how many species of organisms, small organic molecules, and larger polymers would be found? No one knows. Some guess that the number of species in the biosphere is on the order of 100 million. To put us in our taxonomic place, there are more species of insects than of all the vertebrates together. How many kinds of small molecules? Again, no one knows, but we have some clues. By now, huge indexes of organic molecules have been compiled. My friend David Weininger, founder of Daylight Chemicals, which carries out sophisticated analyses of the structure of organic molecules by computer, tells me that on the order of 10 million different organic molecular structures are listed worldwide. Many of these compounds have been synthesized by pharmaceutical and chemical firms. But since it is likely that very many different small molecules in the vast diversity of organisms have never been isolated and characterized, it is a reasonable bet that the naturally occurring organic molecular diversity of the biosphere, limiting our attention to small-size molecules having no more than perhaps 100 carbon atoms, is likely to be 10 million or greater.

How many kinds of large polymers might exist? If we limit ourselves to proteins, we can make crude, very crude, estimates. The human genome—that is, the genes in each cell in your body—encodes about 100,000 proteins. If we make the oversimple assumption that each of the estimated 100 million species on earth makes entirely different proteins, then the protein diversity of the biosphere would be on the order of 100,000 × 100,000,000, or about 10 trillion. Of course, the proteins in related species are highly similar, so this estimate is both crude and probably an overestimate. Still, we will not be too far off if we guess that the biosphere harbors about 1 trillion different proteins.

Ten million small organic molecules and 1 trillion proteins? Nothing like that was around the neighborhood 4 billion years ago. Where did all this diversity come from?

We need new laws. Even controversial candidate laws may help. In this chapter, I will try to persuade you that, in a precise and completely nonmystical sense, the biosphere as a whole may be collectively autocatalytic and—somewhat like the nuclear chain reaction—collectively supracritical, collectively catalyzing the exploding diversity of organic molecules we see.

But while the biosphere as a whole is supracritical, like a mass of fissioning atomic nuclei, the individual cells that make up the biosphere must be subcritical; otherwise, the internal cellular explosion of diversity would be lethal. This, I will try to persuade you, is the source of the creative tension that brings about the ever-increasing diversity of the biosphere. In that tension we may find new law. I want to explore the possibility that this tension drives communities of cells to balance on a phase transition between the subcritical and supracritical regimes, which in turn drives the creation of molecular novelty in the biosphere.

Biological Explosions

If life crystallized as collectively autocatalytic sets, if the fire of life's catalytic closure was kindled by a phase transition in chemical reaction graphs in which a giant web of catalyzed reactions sprang into existence sudden and pregnant, then the first life was already supracritical, already exploding. If so, life has always struggled to control this explosion.

Recall from Chapter 3 our toy model: ten thousand buttons on the floor, you persistently tying randomly chosen pairs of buttons together with threads, pausing every now and then to hoist a button and check how many buttons you haul up in a connected cluster. And recall the phase transition when the ratio of buttons to threads passes the critical value of 0.5—all of a sudden a giant connected cluster, a giant component in the random graph, forms. You lift perhaps 8,000 buttons when you lift one.

This is not yet supracritical behavior. The threads only connect the buttons. The act of connecting the buttons does not in itself create still more buttons and still more threads. But what if it did? Buttons and threads might blossom all over your floor, spill out the window, and smother the neighborhood in a wild, topsy proliferation.

Buttons and threads cannot do such odd things, but chemicals and chemical reactions can. Chemicals can be catalysts that act on other chemical substrates to create still further chemical products. Those novel chemical products can catalyze still further reactions involving themselves and all the original molecules to create still further molecules. These additional new molecules afford still further new reactions with themselves and all the older molecules as substrates, and all the molecules around may serve as catalysts for any of these newly available reactions. Buttons and threads cannot, but chemicals and their reactions can spill out the window, flood the neighborhood, create life, and fill the biosphere.

This explosion of molecular species is what I mean by supracritical behavior. Supracriticality is already present in our model of the emergence of life as collectively autocatalytic sets. Recall that we studied this question by considering a set of polymers, perhaps small proteins, perhaps RNA molecules. We chose these molecules because they can simultaneously serve as substrates for reactions and as catalysts of those same kinds of reactions. This is the important point: molecules can be both substrates and catalysts.

Recall that we used a very simple model for which polymer acts as a catalyst for which reaction; any polymer, we said, has a fixed probability of one in a million, say, to be able to act as a catalyst for any given reaction. When the diversity of molecular species in our pot reached a critical level, the molecules catalyzed so many different reactions that a giant web of catalyzed reactions emerged. Within that giant web were collectively autocatalytic sets of molecules—chemical networks that could sustain themselves.

But that is not the whole story. The next part of the truth, as I know it to tell, is that such a system can keep exploding in diversity, creating more and more kinds of molecules until limited by other factors, such as the supply of food molecules, the concentrations of the molecules, the availability of energy, and so forth. Such a system, at least in the computer simulations we do in silico, is supracritical.

We have also already seen subcritical behavior in our model of the origin of life as the crystallization of collectively autocatalytic sets: if the probability that any polymer catalyzes any reaction is, say, one in a million, and if the diversity of polymers in the system is too small, then no reactions, or almost no reactions, are catalyzed, hardly any new molecules form, and the flurry of molecular novelty quickly dies out.

One easy way to study this process is to imagine "feeding" our hypothetical chemical stew with a constant supply of simple food molecules. Recall that in constructing an autocatalytic set we used the monomers A and B, and the four possible dimers: AA, AB, BA, and BB (Figure 3.7). These molecules combined to form more complex molecules, and when a threshold of complexity was reached, autocatalytic sets crystallized from the confusion. To see at what point such a system becomes supracritical, creating an explosion of new molecules, one might "tune" the diversity of the "food set" to include all the possible trimers—AAA, AAB, ABB . . .—or all the tetramers, and so forth. In addition, one might tune the probability that any polymer can act as a catalyst for any of the possible ligation, or cleavage, reactions among the polymer substrates. Figure 6.1 shows what happens, plotting the length of the longest kind of food molecule—dimer, trimer, tetramer, and so forth—on the x-axis, and the probability that any polymer can catalyze a reaction on the y-axis.

What happens is just what you will now suspect. There is a phase transition line in Figure 6.1 separating two regimes. When the probability of catalysis is low, or the diversity of kinds of food molecules is low,

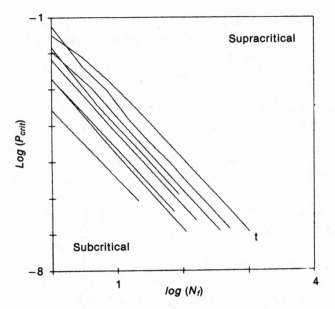

Figure 6.1 *A phase transition. Using logarithmic scales, the number of kinds of molecules in an autocatalytic network is plotted against the probability that any molecule catalyzes any specific reaction. Probability of catalysis is higher at the top of the y-axis than at the bottom. The line labeled* t *predicts a critical phase-transition curve separating supracritical behavior for larger diversities of molecules or higher probability of catalysis, and subcritical behavior for lower diversities or lower probability of catalysis. Other lines paralleling the curve labeled* t *are numerical simulations with polymers using two to 20 kinds of monomers. As the kinds of monomers increases, the numerical curves shift farther away, but remain parallel to the predicted* t *curve.*

or both, the generation of novel kinds of molecules soon dwindles to nothing. Behavior is subcritical. By contrast, when the probability of catalysis is sufficiently high, or the diversity of kinds of food molecules is sufficiently high, or both, the system is supracritical and generates an explosion of new kinds of molecules that in turn catalyze the formation of still further new kinds of molecules that in turn beget still new kinds of molecules. Our chemical buttons and threads here create more buttons and threads ad infinitum.

Supracritical Soup

It is one thing to talk about supracritical reaction systems blasting off into the outer space of chemical creativity, but all confined to a computer model. It is quite another thing to fathom what might go on in a real chemical system. Let us turn to creating a supracritical soup—wet and filled with more diversity than anything Campbells ever dared to try. If we can do it in a vat, you can bet that nature can do it in the fullness of time. Remember, we are after the laws that govern the explosion of molecular diversity and complexity of the biosphere.

To proceed, we have to think about the kinds of molecules in our pot and all the reactions they can undergo. And we need a set of enzymes to catalyze those reactions, assigning each a probability that it will actually be able to catalyze any given reaction. Then we can ask whether we expect that our real soup of molecules will actually be supracritical or subcritical.

The first step is to consider how many kinds of molecules and kinds of reactions can be created with our recipe. This is, in fact, a very difficult question. Organic molecules are not simply straight chains of linked atoms; they are often complex structures with branches and interlinked rings. The chemical formula of a molecule is given by listing the number of copies of each type of atom. For example, $C_5H_{12}O_4S$ has five carbons, 12 hydrogens, four oxygens, and one sulfur atom. It is exceedingly hard to count the number of possible molecules that can be made with this chemical formula. But this much is sure: as the number of atoms per molecule increases, the number of kinds of possible molecules increases extremely rapidly. If I asked you to make all the kinds of molecules possible with, say, 100 atoms, using carbon, nitrogen, oxygen, hydrogen, and sulfur, you would go nuts trying to list the possibilities. It is another of those astronomical numbers.

It is also hard to count the number of types of chemical reactions that might occur, according to quantum mechanics and the laws of chemistry. In general, we can think of reactions in which one substrate converts to one product, two substrates are joined to form one product, one substrate breaks apart to form two products, and two substrates exchange atoms to form two new products. Two-substrate, two-product reactions are very common, and typically involve breaking one or more atoms off one molecule and bonding them onto the second.

While we do not know how many reactions among a complex set of molecules is actually possible, I want to build a crude estimate that will serve our purposes. I am going to estimate that for any two reasonably complex organic molecules, there is at least one legitimate two-substrate, two-product reaction the two can undergo.

This is almost surely an underestimate. Consider small polynucleotides (called oligonucleotides) of the kind discussed in Chapter 2—say, CCCCCCC and GGGGGGG, each with seven nucleotides. In either molecule, any internal bond can be broken and the "right ends" exchanged to create two new molecules. An example would yield CCCGGGG and GGGCCCC. Since each molecule has six bonds, there are 36, rather than merely one, possible two-substrate, two-product reactions between these two substrates.

If any pair of modestly complex organic molecules can undergo at least one two-substrate, two-product reaction, then the number of reactions afforded by a complex mixture of molecular species is at least equal to the number of kinds of *pairs* of organic molecules. Thus if there are 100 organic molecule species, the number of pairs is just 100×100, or 10,000.

The important point is this: if the number of kinds of molecules is N, then the number of kinds of reactions is N squared. N squared increases rapidly as N increases. If there were 10,000 kinds of molecules, there would be about 100 million kinds of two-substrate, two-product reactions among them!

Finally, suppose we want to use proteins as our candidate enzymes. Then we need to think about the probability that each can act as an enzyme to catalyze any one of the possible reactions. If we want to make a real supracritical soup, it turns out to be wise to consider using antibody molecules as our candidate enzymes. This choice is not essential. Everything I am about to describe should hold true if we use other proteins as catalysts. At this point, however, future experiments with supracritical solutions can make use of the surprising fact that antibody molecules, evolved to vanquish invaders, can catalyze reactions. Antibody molecules that do this are called catalytic antibodies, or abzymes. The experimental procedure to generate such abzymes concerns finding antibody molecules that can bind to the transition state of a reaction. Almost 1 in 10 such antibodies can then actually catalyze the reaction itself. From the data on abzymes, it is now pretty clear that the probability that a randomly chosen antibody molecule will catalyze a randomly chosen reaction is something like one in a million. If we guess that the probability is between one in a million and one in a billion, we'll be pretty safe.

Now we are ready to cook up a pot of supracritical soup. Imagine a chemical reaction system in some appropriate vessel. Vary the number of kinds of organic molecules, on the one hand, and the number of kinds of antibody molecules, on the other. Plot antibody diversity on the x-axis, and organic molecular diversity on the y-axis (Figure 6.2). Now consider what will probably occur near the "origin" of this coordinate system, where two organic molecules and one antibody molecule are in the reaction system. The chance that the antibody catalyzes any one of the four possible two-substrate, two-product reactions is only

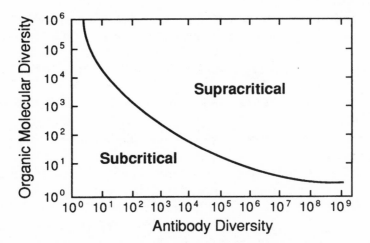

Figure 6.2 *Cooking up a pot of supracritical soup. Using logarithmic scales, the diversity of antibodies, serving as candidate catalysts, is plotted against the diversity of organic molecules serving as substrates and products. A curve denotes the approximate shape of a phase transition between subcritical behavior, below the curve, and supracritical behavior, above the curve.*

one in a billion or so. The chance that any of the four reactions are catalyzed is nearly zero. Hence the formation of new molecular species is almost certainly not catalyzed. The behavior of our soup is subcritical. The soup is dead chicken soup.

Imagine instead that there are 10,000 kinds of organic molecules in the system and 1 million kinds of antibody molecules. Then the number of reactions is 10,000 × 10,000, or 100,000,000. There are 1 million antibodies, each a candidate to catalyze any of the possible 100 million reactions. The chance that any antibody catalyzes any reaction is again, let us say, one in a billion. Thus the expected number of reactions for which antibody catalysts are present in the system is the number of reactions times the number of kinds of candidate catalysts divided by the probability that any given catalyst catalyzes any given reaction. The result is that the expected number of catalyzed reactions is 100,000.

If the system has catalysts for 100,000 reactions, then the initial 10,000 types of organic molecules will rapidly undergo about 100,000 reactions. Common sense tells us that most of the products of these reactions will be entirely new. Thus the organic molecular diversity in the system will rapidly explode from 10,000 to something like 100,000.

Once the system is ignited, it will keep exploding in diversity. The first round of catalyzed reactions yields about 100,000 molecular species, but because of this, the number of potential reactions now explodes to the square of 100,000, yielding 10 billion possible reactions!

The same 1 million antibody molecules are candidates to catalyze these new reactions, and about 10 million of the reactions will find antibody catalysts. Thus in the next round of catalyzed reactions, about 10 million new molecular species will be created. The initial 10,000 species has exploded 1,000-fold, and the process continues, with the diversity of molecular species mushrooming. This explosion of diversity is supracritical behavior.

With a little thought it becomes obvious that, in our xy Cartesian coordinate system, some curve separates subcritical from supracritical behavior (Figure 6.2). If there are few organic molecules and few antibodies, then the system is subcritical. But if the number of kinds of antibody molecules is held fixed at some low number, say 1,000, and the number of kinds of organic molecules is increased, eventually there will be so many possible reactions among these organic molecules that the 1,000 antibody molecules will catalyze reactions leading to a supracritical explosion in organic molecular diversity. Similarly, if the number of kinds of organic molecules is held fixed, say 500, and the diversity of antibody molecules is increased, eventually a diversity will be reached such that the antibodies will catalyze reactions leading to a supracritical explosion. It follows that the critical curve in the xy coordinate system is roughly as shown (Figure 6.2). For diversities of organic molecules and antibodies that are below the critical curve, the system is subcritical. For diversities of organic molecules and antibody molecules above this curve, the system is supracritical.

Supracritical soup. A rich minestrone.

The increasing grounds to suppose that real chemical reaction systems can be supracritical suggest the possibility that the biosphere itself is supracritical. We explore this next, for supracriticality is, I believe, the ultimate wellspring of molecular diversity in the biosphere. Yet that very creativity poses the most profound dangers for the cells that create it, nest within it, are sustained by it, but must survive it. If the biosphere is supracritical, how do cells protect themselves from the molecular chaos that supracriticality implies? The answer, I think, is that cells are now, and always have been, subcritical. But if so, we shall have to explain how a supracritical biosphere can be made of subcritical cells. In our search for big game, we may be on the track of a new biological law.

The Noah Experiment

Sometime in the past day, you ate dinner. Knife and fork, chopsticks, hungry hands—it is utterly familiar that we pick up food, bring it to our mouths, chew, and swallow. The food we eat is broken down by digestion into simple small molecules, which we absorb and build up again into our own kinds of complex molecules. Why bother with this form of taking in matter and energy? Why not just walk up to the Caesar-garnished lettuce leaf, cry "Be mine!" and fuse with it? Why do we not simply fuse our cells to those of the spinach soufflé, mixing their metabolic wealth with our own? Why, in short, bother with the fumbling uncertainty of digestion breaking apart molecules only to build them up again?

That we eat our meals rather than fusing with them marks, I believe, a profound fact. The biosphere itself is supracritical. Our cells are just subcritical. Were we to fuse with the salad, the molecular diversity this fusion would engender within our cells would unleash a cataclysmic supracritical explosion. The explosion of molecular novelty would soon be lethal to the unhappy cells harboring the explosion. The fact that we eat is not an accident, one of many conceivable methods evolution might have alighted on to get new molecules into our metabolic webs. Eating and digestion, I suspect, reflect our need to protect ourselves from the supracritical molecular diversity of the biosphere.

It is time to carry out a thought experiment—the Noah's Vessel experiment. Take two of each species—fly, flea, sweet pea, moss, manta ray—with some effort to normalize for big versus small; add a small scoop of hippo-stuff popped into our vessel for every fern frond. A hundred million species, two by two. Then, with the sophistication of any well-trained biochemist, grind them all up. A pestle will do. Grind them up, dispersing the cell membranes, dispersing the intracellular organelle membranes, releasing the saturated sap of each life to mingle with the pregnant sap of all life.

What will happen? Ten million small organic molecules will mingle with 1 trillion or so proteins, all in a rich soup, nuzzling together. The 10 million organic molecules afford some 100 trillion possible reactions! Each of the 1 trillion proteins, well wrought by evolution for its separate function in some cell, is nevertheless laden with nooks and crannies. Each protein, by inadvertent happenstance, harbors clefts and bumps that, like the many molecular locks of the antibody repetoire, just might bind the transition states of one or more of the 100 trillion possible reactions. Suppose that the probability that any protein has a binding site allowing it to act as a catalyst for any randomly chosen reaction is even lower than one in a billion—say, one in a trillion. How many reactions will be catalyzed? One hundred trillion reactions times 10 trillion proteins divided by 1 trillion is, well, 10^{15}, or 1 quadrillion. This is more than the total number of possible kinds of reactions in the system. In effect, each of the 100 trillion reactions would find 10 protein catalysts! A hundred trillion products would be formed from the 10 million kinds of starting organic molecules. A vast explosion of diversity would carom off the perplexed walls of Noah's groaning vessel.

Our estimates could be wrong by orders of magnitude (powers of 10), and our broad conclusion would still hold: the biosphere is supracritical. And that supracriticality is, I believe, the very foundation for the increase in molecular diversity and complexity over the vast range of time since the first living forms graced the earth. If the earth is supracritical, then this must be revelatory of the laws of complexity—of the ways the nonequilibrium universe creates, and ultimately created us.

But if the biosphere as a whole is supracritical, what of cells? How would they survive, how could they persist and evolve in such a volatile world?

Consider a human cell, say a liver cell. About 100,000 proteins are encoded in the human genome. No cell expresses all these at once, but suppose your liver cell did. If you look at a chart of metabolism (for example, Figure 6.3), you will see about 700 to 1,000 small organic molecules listed. These undergo a variety of reactions, the traffic that constitutes the metabolism of the cell. Now it is important that the proteins in your liver cell have evolved to catalyze the desired reactions and to *not* catalyze undesired side reactions. Nevertheless, as noted earlier, each of the 100,000 proteins in a liver cell has a host of nooks and crannies that might bind novel transition states and catalyze novel reactions.

Let us conceive of a new experiment. Obtain a novel organic molecule, call it Q. Inject Q into the waiting liver cell. What will happen? Well, let us admit that each of the 1,000 organic molecules might form one member with Q of a two-substrate reaction. Thus let us guess that the injection of Q affords about 1,000 novel reactions. To be conservative, let us guess that the probability that any one of your 100,000 proteins catalyzes any of these novel reactions is one in a billion. How many reactions will find catalysts? The answer, as before, is given by the product of the number of reactions times the number of candidate enzymes, divided by the probability of catalysis. Thus $1,000 \times 100,000/1,000,000,000 = 0.1$. In short, the chance that even one of the 1,000 reactions involving Q will be catalyzed by some protein is on the order of 1/10. But this means that your liver cell is, in fact, subcritical. It seems intuitively clear that no avalanche of reactions forming novel molecules will be unleashed by the injection of Q into the cell. Q is unlikely to form a novel product. Even if it did form such a product, say R, R is equally unlikely to afford a reaction that is catalyzed to form a still further novel molecule, S. Any effect of Q quickly fizzles out.

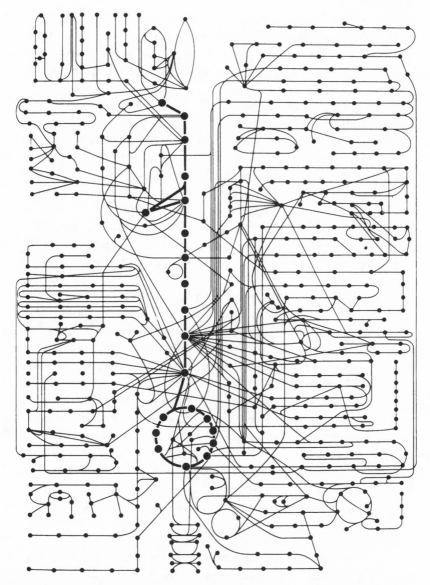

Figure 6.3 *Cellular traffic. In this chart of intermediate metabolism in humans, 700 or so small molecules interact. The nodes correspond to metabolites; the edges, to chemical transformations.*

Indeed, the intuition is correct. A mathematical theory called branching processes shows that if the expected number of "progeny" of a "parent" is greater than 1.0, then the lineage is expected to branch and increase in diversity indefinitely. If the expected number of progeny is less than 1.0, then the lineage is expected to die out. (Chain reactions in nuclear explosions are just those in which the number of daughter neutrons created by a single hit by one neutron into a uranium nucleus is greater than 1.0.) Since the expected number of progeny derived from the parent molecule, Q, is just 0.1, much less than 1.0, our crude calculation shows that our liver cells are subcritical.

Should we believe the details of this back-of-the-envelope calculation? Not at all. Too many rough guesses enter the simple formula. I do not really know how many novel reactions are afforded by the injection of Q, nor do I know the real number of protein and other candidate enzymes any one of our cells may actually provide. More critically, no one knows the true probability that a randomly chosen protein in a liver cell will catalyze a randomly chosen reaction. That probability may be as high as, say, 1 in 100,000, or as low as 1 in 1 trillion. But the crude calculation suffices to show this: if the biosphere as a whole is supracritical, cells are probably somewhere below the subcritical–supracritical boundary.

If cells are subcritical, this must be a fact of enormous importance. It must be a candidate for the biological laws we seek. Suppose cells were actually supracritical. Then the injection of a novel molecule Q would unleash a cascade of molecular novelty: Q, R, S. The cascade would propagate, since each novel molecule in turn would afford reactions creating still further novel molecules. Almost certainly, many of these novel molecules would disrupt the homeostatic molecular coordination within the cells, leading to cell death. In short, supracriticality in cells would soon be lethal. What protection might cells have evolved? Cells might use carefully wrought membranes to exclude all "foreign" molecules, or they might develop an immune system. But the simplest defense, surely, is to remain subcritical.

We may be discovering a universal in biology, a new law: if our cells are subcritical, then, presumably, so too are all cells—bacteria, bracken, fern, bird, man. Throughout the supracritical explosion of the biosphere, cells since the Paleozoic, cells since the start, cells since 3.45 billion years ago must have remained subcritical. If so, then this subcritical–supracritical boundary must have always set an upper limit on the molecular diversity that can be housed within one cell. A limit exists, then, on the molecular complexity of the cell.

And yet while cells each in themselves are subcritical, they interact to form communities such that the globe as a whole is supracritical and eases ever upward, creating the molecular diversity just out the window. How did this delicate balance arise? We turn now to ask how, if cells are subcritical, the biosphere driven by cells comes to be supracritical. The answer may be that communities of cells might evolve to the subcritical–supracritical boundary.

Avalanches of Novelty

We do not live alone. I am not referring to our bustling conjunction with one another on this increasingly crowded planet. I am referring to our intestines. A menagerie of single-celled organisms—most happily, bacteria—flourish as you feed. Nontrivial, these, for their metabolic activities are critical to our health. These bacteria and other creatures form little ecosystems. But there are many such relatively isolated ecosystems, ranging from the communities clustered around the hydrothermal hot vents on the sea floor, to communities crouched along the edges of hot springs in the far fields of Iceland, to complex communities of bacteria and algae that mimic the communities that formed the ancient stromatoliths.

Can a local patch of an ecosystem in the intestines or elsewhere be supracritical? If so, that ecosystem itself would be able to generate an explosion of molecular diversity. Let us reinterpret Figure 6.2, which shows antibody molecule diversity on the x-axis, and small molecule diversity on the y-axis. Instead, we plot bacterial species diversity on the x-axis, and the diversity of novel molecules imposed on the community on the y-axis (Figure 6.4). As the diversity of bacterial species increases,

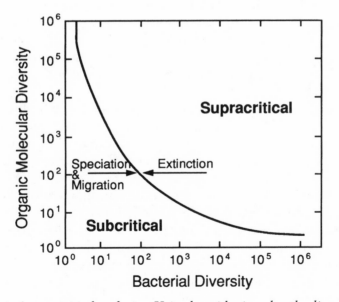

Figure 6.4 *A supracritical explosion. Using logarithmic scales, the diversity of species in a hypothetical bacterial ecosystem is plotted against the diversity of organic molecules added from outside. Again, a curve separates subcritical behavior from supracritical behavior. Arrows pointing toward the critical phase-transition curve show hypothetical evolution of ecosystem diversity to the boundary.*

the total diversity of polymers present in the ecosystem increases. Similarly, on the y-axis, we can consider the addition of increasing diversities of novel molecules to the ecosystem.

What will happen? Consider our old friend, Q. If Q insinuates itself into some cell, it will be either sequestered in that cell or resecreted from the cell otherwise unchanged—or Q might undergo a reaction forming a novel molecule, R. Now R might be sequestered in that cell, or secreted or released from that cell. But what will R do? If secreted, it might find its way into some other cell, perhaps of a different bacterial species. And the story repeats, for R may engender the formation of S.

Consider Figure 6.4. Impose a fixed level of exogenously supplied novel molecules on a series of bacterial ecosystems of increasing species diversity. At some critical diversity and above, one would expect that the ecosystem as a whole will be supracritical. Molecular novelty will be amplified by the clustered community of cells. Or instead, hold the diversity of species low, but increase the diversity of exogenously supplied novel molecules. At and above some critical diversity of molecules, the ecosystem should again become supracritical. Thus, once again, some kind of curve should separate subcritical and supracritical regimes.

In short, pack diverse cells together and we approximate the Noah's Vessel experiment, but with the cell membranes slowing the story a bit. Cell to cell, molecules travel, meet, transform. Cell to cell, molecules create their own story of becoming, using the cell's machinery to speed the tale. Enzymes can only ratify organic chemistry. The supracritical explosion of the biosphere is immanent in the combinatorial character of chemistry itself, propelled forward by life itself, the finest triumph of supracriticality.

Clearly though, each ecosystem must somehow rein in this explosive tendency. The exploding molecular diversity will be toxic to members of the ecosystem. What might we expect to occur? Will some new balance, some compromise between subcriticality and supracriticality, be struck?

Suppose the little ecosystem happens to be supracritical. Some hapless species of bacteria will have the unfortunate capacity to transform the novel molecular types impinging on it into a virulent internal toxin. The toxin will kill that cell. One must imagine that all of the same type of bacterial cells in the ecosystem will go extinct and pass to bacterial heaven. This loss, tragic in its way, this extinction of one species of bacteria from the little ecosystem, will have a simple consequence. The extinction will *lower* the diversity of species present in the ecosystem. Since the community is supracritical, a variety of novel molecules will be persistently formed, but, in turn, these will drive a succession of such extinction events. In turn, the extinction events should tend to lower the species diversity of the ecosystem from the supracritical regime, toward the subcritical regime (Figure 6.4).

Suppose instead that the ecosystem is subcritical. Too few bacterial types, harmonious in their metabolic tradings, mutualisms, and minor competitive strife, wander the metabolic stage to create much molecular novelty. But from offstage left enters a succession of immigrant species, newcomers from across the tracks. Worse, some of the old-timers, too bored to remain ever the same genetically, and given the whatnots of random mutations and genetic drift, evolve novel large polymers—DNA, RNA, and proteins that may catalyze new reactions. This persistent immigration and speciation increase the diversity of bacterial species in the ecosystem, driving the ecosystem from the subcritical toward the supracritical regime (Figure 6.4).

So we have a candidate for a biological law. The twin pressures from the supracritical regime toward the subcritical regime and from the subcritical regime toward the supracritical regime may meet in the middle, at the boundary between subcritical and supracritical behavior. The temptation of this thought is obvious. One cannot help but wonder whether local ecosystems evolve to the subcritical–supracritical boundary, evolve to that boundary and then remain poised there ever after, held by the pressure of opportunity if subcritical, and by lethality if supracritical. Ecosystems may evolve to the subcritical–supracritical boundary!

A further conclusion: while each ecosystem is itself on the boundary between sub- and supracriticality, by trading their stuff they collectively produce a supracritical biosphere, one that inexorably becomes more complex.

If these hypotheses are correct, the behavior of the biosphere is analogous to that of a nuclear reaction, not a nuclear explosion. In a nuclear explosion, the chain reaction "runs away" autocatalytically. But in a nuclear reactor, carbon rods absorb the excess neutrons and ensure that the branching probability of the chain reaction stays at or below the critical value of 1.0. Useful energy is obtained, not a massive explosion. If local ecosystems evolve to the subcritical–supracritical boundary, then the branching probability of molecular novelty should be near 1.0, leading to small and large bursts or avalanches of molecular novelty. Critical value branching processes give rise to a power-law distribution of avalanches like the one encountered before in the sandpile model of Per Bak, Chao Tang, and Kurt Wiesenfeld. But, as in the controlled nuclear reactor, these avalanches of novelty each eventually die out. The next avalanche, small or large, will be triggered by the next fluctuation that brings in some novel molecule Q. Like the sandpile grains of sand, where each may unleash a small or large avalanche, the poised ecosystems will create small and large bursts of molecular novelty.

If this view is correct, the poised biosphere creeps upward in a controlled generation of molecular diversity, not a vast wild explosion. The wild explosion would occur if the biosphere were the Noah's Vessel experiment, with all cell membranes disrupted. Cell membranes block many molecular interactions and hence block a supracritical explosion, just as carbon rods in a reactor absorb neutrons and hence block collisions of those neutrons with nuclei, which would lead to a supracritical chain reaction.

Do we yet know that local ecosystems evolve to the subcritical–supracritical boundary? Not at all. Is it plausible? I think so, with caveats. We have not, for example, considered explicitly the fact that as molecular diversity explodes in a supracritical system, the concentrations of products can decrease, slowing the forward rates of subsequent reactions. The concentration effects are surely important, but seem unlikely to alter the basic conclusions. Can we test the hypothesis, using modern tools such as mass spectroscopy and others to discriminate the generation of molecular novelty? I believe so.

And if so? If local ecosystems are metabolically poised at the subcritical–supracritical boundary, while the biosphere as a whole is supracritical? Then what a new tale we tell, of life cooperating to beget ever new kinds of molecules, and a biosphere where local ecosystems are poised at the boundary, but have collectively crept slowly upward in total diversity by the supracritical character of the whole planet. The whole biosphere is broadly collectively autocatalytic, catalyzing its own maintenance and ongoing molecular exploration. We tell a tale of the organisms of the biosphere collectively catalyzing a sustained drive toward increased molecular diversity and complexity. We tell a tale of cells coping with the happy and distressing consequences of their own catalytic activities, ever forced to transform or die, doing both. As with the sandpile, the actors can never know if their next step, their latest experiment, will unleash a small trickle of molecular change or an all-engulfing avalanche. And the avalanche might be one of creativity or one of extinction.

We tell a tale of cells, of life-forms, always hitching up their pants, always putting on their galoshes, always doing their local, level best. We tell a tale of humble wisdom, for each creature evolving as best it can, inevitably creates the conditions of its own ultimate elimination.

We, humbly at home, strutting and fretting our hour, part of the pageant, part of the unfolding census counted at one brief moment of history by Noah.

Chapter 7

The Promised
Land

We call ourselves by many names: *Homo sapiens,* man the wise; *Homo habilis,* man the able, the toolmaker; and, perhaps the most appropriate, *Homo ludens,* man the playful. Each aspect makes its contributions to science. There is no one way, for our exploration depends on all three: wisdom to choose the ill-lit path; techniques to find the answers; but always, always beneath the surface, playfulness. "Theory is the free invention of the human mind," said Einstein, one whose wisdom and technique were almost beyond our ken, but at base, one who could play with the utmost freedom.

It is this, the science of *Homo ludens,* that expresses the most precious creative core of humanity. Here science is art. What honor and joy to seek the laws. Einstein sought the secrets of the Old One. We, although lesser, hope for no less. And sometimes, more often than should be the case, the playful joy of practicing science unexpectedly brings forth news of a promised land beyond the Sinai—technologies that will change the way we live.

Ill-planned or not, happenstance or not, we are passing over a new threshold in humanity's history. "Passing" understates the case; rather, we are racing over a vast threshold of untold promise, and perhaps untold danger. For the first time in human history, we can challenge the molecular diversity accumulated in the biosphere. We can create vast ranges of novel molecules. Among these we may find fabulous new medicines. We may also find dangerous new toxins. Faust made his deal; Adam and Eve ate of the tree of knowledge. We cannot keep from going forward. We cannot shrink from our vision; yet we cannot foresee the consequences. So it was too when *Homo habilis* hefted a hafted stone with conquest blossoming in his heart. Neither he nor we can foresee the consequences of the ripples we must make. We live on a sandpile; we ourselves create the sand; we ourselves step across the sandpile, shedding avalanches at each pace. Gone is the presumption that we can predict more than a bit.

What is this new Faustian bargain? We saw in Chapter 6 that the biosphere holds something like 10 million small organic molecules and perhaps 10 trillion proteins and other polymers. We saw evidence that the biosphere may be balanced on a subcritical–supracritical boundary, setting off small and large bursts of molecular creativity. We found ourselves wondering if the linked metabolisms of the organisms spread across the earth might not be the most complex chemical factory in a few dozen megaparsecs of space.

The revolution in biotechnology, the mystery and mastery of cloning the genestuff itself, tells us that we can blast forth a diversity of molecular forms such as have probably never been assembled in one place at one time in the 15-billion-year unfolding of the cosmos. New medicines and new toxins understate the case. We know not what we are about to do with our molecules of sand piled high with pride.

Applied Molecular Evolution

We now presume to surpass the rendered wisdom of 3.45 billion years of molecular evolution with a new sandbox of a field called applied molecular evolution. We, many of us now—molecular biologists, chemists, biologists, biotechnology companies, the pharmaceutical industry—are about to generate molecular novelty at rates and on scales never before achieved.

Before launching our journey into biotechnology's new frontier, already being called by some its second birth, it is worth reviewing the now familiar but stunning magic of current gene splicing and gene cloning, with its marvelous medical promises. As we have seen, each cell in your body houses DNA that encodes about 100,000 proteins. These proteins play structural and functional roles in the lives of your cells. Since the 1970s, it has become clear how to snip out sequences of human genes, snap the fragments into the DNA of viruses or other carriers, and use them to infect bacterial or other cells. In this manner, these host cells are tricked into manufacturing the protein encoded by the human gene. This entire process is lumped under the term "gene cloning."

The medical potential of gene cloning was not long lost on the scientific and the business community. Many diseases are caused because the DNA sequence that codes for an important protein has become altered by a random mutation. By cloning the normal version of a human gene, we can manufacture the normal protein product and use it to treat the disease. Perhaps the most commercially successful product is EPO, the erythropoetin hormone, which fights anemia because it stimulates the production of red blood cells. Erythropoetin is a small protein, called a peptide. Amgen won a major legal tussle with another company, and now is doing very well marketing this product of 3.45 billion years of evolutionary research. Estimates of the medical uses for cloned human genes, in sheer monetary terms, are extremely large. Given the fraction of gross national product devoted to health care in the United States and elsewhere in the developed and developing world, and needs for medical help with a universally aging population, it is no wonder Wall Street rushed to invest in this technology with such enthusiasm. While few products have yet reached market, and investor enthusiasm waxes and wanes, billions of dollars have been raised and invested in this particular vision of the promised land.

Moreover, the human genome project is well under way. Since the genome encodes all the proteins that occur in us, the time is not distant when all the normal proteins of humans and their patterns of variation in human populations will be known. This promises a vast wealth of knowledge, for each of those proteins can malfunction. Knowing the proteins, we should be able to devise appropriate therapies, from increasingly subtle medicines to the replacement of faulty genes through gene therapy.

What do I mean by subtle medicines? We have already discussed the genomic regulatory network in each cell that coordinates gene activities in ontogeny, the unfolding of the embryo. Our lightbulb network models have illustrated that a slight perturbation of one bulb's activity can cause cascades of alterations to carom through the network. Our drugs impinge on our tissues because they bind to molecules in our cells; by binding, the drug unleashes a cascade of alterations. Among these alterations, some are beneficial, and others are the familiar unwanted side effects listed in very small print on our prescriptions, if they are known at all. By subtle, I mean that we want precise molecular chisels to create the therapeutic response and cleave off the side effects. Where will these subtle medicines come from? From the enormous diversity of new kinds of molecules, DNA, RNA, proteins, and small molecules we now can create.

Traditional biotechnology has sought to clone human genes and their proteins. But why should we believe that the only source of useful proteins are those already in us? Let us just think for a moment how many possible proteins there are, limiting our attention to proteins with only

100 amino acids. (Most biological proteins have several hundred amino acids.) Biological proteins use 20 kinds of amino acids—glycine, alanine, lysine, arginine, and so forth. A protein is a linear sequence of these. Picture 20 colors of beads. A protein of 100 amino acids is like a string of 100 beads. The number of possible strings is just the number of types of beads, here 20, multiplied times itself 100 times. Now, 20^{100} is roughly equal to 10^{120}, or a 1 with 120 zeros after it. Even in these days of vast federal deficits, 10^{120} is a really big number. The estimated number of hydrogen molecules in the entire universe is 10^{60}. So the number of possible proteins of length 100 is equal to the square of the number of hydrogen molecules in the universe.

We might suppose that, through natural selection, nature has tried out this stunning number of combinations, rejecting all but the small subset that have found their way into earthly life—that there is no reason for us to look any farther than our own genomes. But this, on reflection, is not reasonable. Recall the calculations in Chapter 2 about the chance that a cell, made up of a certain set of proteins, would spring forth fully assembled. Robert Shapiro estimated the number of trials that might have occurred in all the oceans ever dreamed of by Columbus. I think this part of his argument is reasonable. Assuming that a "trial" occurs in a volume of one cubic micron and takes one microsecond, Shapiro calculated that enough time has elapsed since the earth was born to carry out 10^{51} trials, or less. If a new protein were tried in each trial, then only 10^{51} of the 10^{120} possible proteins of length 100 can have been tried in the history of the earth. Thus only a tiny portion of the total diversity of such proteins has ever existed on the earth! Life has explored only an infinitesimal fraction of the possible proteins.

If such a tiny fraction of the potential diversity of proteins of length 100 have ever felt the sun's warmth, then there is plenty of room for human explorers to roam. Evolution can have sampled only the tiniest reaches of "protein space." And since selection tends to stick with the useful forms it finds, evolution's search has probably been even more restrictive.

Applied molecular evolution is based on the fundamental recognition that the near and outer reaches of protein space, DNA space, or RNA space are fully likely to harbor novel molecules of enormous practical benefit. The central idea is simple: generate billions, trillions, billions of trillions of random DNA or RNA or protein sequences, and learn to search this enormous, staggering diversity of molecular forms for the useful molecules we need. For better or worse, we will find fabulous new medicines. For better or worse, we will find frightful new toxins. For better or worse—so it has always been.

I like to call this new approach to drug discovery applied molecular evolution. Sidney Brenner, one of molecular biology's heroes, a discoverer of the triplet genetic code, and source of many slightly scandalous comments, likes to call this new era of drug discovery by a simpler catchphrase: "irrational drug design."

Make enough molecules and select something that works. If need be, improve it. Why not? That's how Mother Nature does it. Thanks to the brilliance of an increasing host of molecular biologists and chemists, this new idea is becoming established; indeed, it is flourishing. At present, some small biotechnology companies are examining peptide, DNA, and RNA "libraries" with diversities in the trillions for useful molecules.

A trillion different DNA, RNA, or protein molecules? Sounds hard, but it is actually relatively easy. I may have made the first such library with about a billion different random DNA sequences, each snapped inside the genome of a billion different copies of a particular virus called lambda. If you want to rush out and make such a library, here is one way: ask a friend with a DNA-synthesizing machine to build up DNA sequences with 100 nucleotides bonded together. Normally, people want to make trillions of copies of the very same 100-nucleotide DNA sequence. But it is no trick to make trillions of different, random, 100-nucleotide DNA sequences at the same time. If you add the proper few nucleotides at the beginning and the end, creating specific sites, the resulting library of molecules is ready to be snapped into place in a viral or other host. To do so, the host's own double-stranded DNA has to be cut with an enzyme to create two single-stranded "ends," each of which terminates in a few nucleotides that match up (through Watson–Crick base pairing) with the special single-stranded terminal nucleotides you built into your library of DNA sequences. Mix your library of trillions of random DNA molecules with trillions of copies of the DNA from the virus whose "ends" are exposed in a tiny tube, add a few well-known enzymes to glue your DNA sequences to the viral DNA, wait a few hours, and . . .

And if you are lucky, you have a library of a trillion different random DNA molecules cloned into a trillion different copies of some lucky virus. Now all you have to do is search among these trillion random genes and see if you can find what you want: sequences that carry out a certain biochemical task.

What kinds of uses can you hope for? New drugs, new vaccines, new enzymes, new biosensors. Ultimately, perhaps, experimental creation of collectively autocatalytic sets, hence new life. I'll sketch these possibilities later.

But how, among billions of more or less random molecules, could you find the useful ones? Some needle-in-the-haystack problem, this— or so it seems at first.

One way to find the needle rests on a very simple idea: if you have a key and would like to create a copy, first make an impression of the key, in clay perhaps, thereby producing the "negative" of the key, its "shape complement." Now use the clay mold either to pour a new molten key or, lacking that, to match any other key in a large library of keys with the pattern. If a second key fits the mold, it looks like the first key. Probably it will open the same lock.

Recall that an antibody molecule binds the corresponding antigen rather like a lock fits a key. Suppose we wish to find a peptide that has a shape similar to some hormone, say insulin. We start by immunizing a rabbit with human insulin. Its immune system reacts to this invader by generating antibody molecules. Think of the insulin as the first key and the antibody molecule as a lock that fits the antigen key. We can take any one of these antibody molecules and clone it to obtain many identical copies. Now throw away the insulin and carry out the following trick: generate about 100 million more or less random DNA or RNA sequences, clone them into a viral host, and express the more or less random peptides or polypeptides this library of genes encodes. (Actually, there are a variety of ways to generate billions of random peptides.) With the library of 100 million peptides in hand, expose all of these simultaneously to the identical copies of the antibody molecule lock. Any peptides that are bound by the antibody molecule must have shapes similar to that of insulin. Thus any such peptides are second keys and are candidates to mimic the activities of the insulin hormone.

We need not use an antibody that binds to the hormone itself. If we have the receptor for insulin or some other hormone—the molecule the hormone must bind to in order to carry out its work—we can use the receptor as the "lock" to fish out a second molecular key fitting that lock, hence mimicking the initial hormone. In each case, the second molecular key is a candidate drug, a candidate therapy that may traverse the hurdles and mature into a therapy you may someday use in the treatment of a hormonal dysfunction.

Nor are we limited to finding molecules that mimic a hormone. Applied molecular evolution should soon be capable of finding new molecules binding to, mimicking, enhancing, or inhibiting the activities of almost any peptide or protein in the body. This means chiseling new therapies for many diseases associated with dysfunctional proteins. For example, the technique can be used to seek peptides that mimic the hundreds or thousands of signaling molecules that bind to any of the hundreds or thousands of specific receptors in the body. The nervous system is but one case. Hundreds of different types of receptors have already been located in the human brain alone, and the list is growing rapidly. These receptors react to neurotransmitters such as serotonin, acetylcholine, adrenalin, and hundreds of others. The capacity to find novel molecules that bind selectively to specific receptors or bind to enzymes that act on neurotransmitters promises very powerful and useful new therapies for nerve system and emotional disorders. Prozac, a highly useful antidepressant, acts by inhibiting an enzyme that breaks down serotonin. Prozac is just a hint of the kinds of molecular precision instruments we will acquire in the therapies coming over the horizon.

Key–lock–key. If this simple idea may bring us a cornucopia of drugs in the near future, the potential for the generation of novel vaccines may be among the most promising prospects of the new promised land.

There are two main established ways to make vaccines. Suppose a virus, like the polio virus, causes a disease. One classic means to create a vaccine is to obtain dead virus and use it. Alternatively, mutants of the virus are found, or created, that attenuate the virus, so that it can confer immunity without causing the disease. In either case, the virus material is injected and your immune system mounts an immune response to ward off future attacks by the same virus. The danger is obvious. What if some of the killed viruses were not quite dead? What if the attenuated virus accumulates "revertant" mutations, which return it to full activity? The newer procedure to make vaccines attempts to circumvent this problem by using genetic engineering to clone the genes coding for the proteins on the surface of the virus that are recognized by the immune system. Then these proteins themselves are injected as the vaccine. The patient never receives the virus itself.

This new procedure can work brilliantly. However, even it has problems. First, one has to know the cause of the disease, the virus or other pathogen. Second, the pathogen must contain proteins that can stimulate an immune response that blocks the pathogen. Third, one has to obtain a supply of that protein—typically by growing the pathogen in the laboratory. Fourth, one has to clone the gene encoding the pathogen and actually make the vaccine. The fourth step is typically feasible if the first three can be carried out.

The key–lock–key procedure and applied molecular evolution suggest a completely different approach to vaccine development, which may find very powerful applications. Suppose, in the example, rather than starting with insulin, I had started with the hepatitis virus and immunized a rabbit. I would obtain antibody molecules directed against specific molecular features, called epitopes, on the hepatitis virus. With such antibody locks in hand, I could search a library of more or less random peptides for ones that bind the antibody and pull out second keys, which must look like the epitope on the hepatitis virus. But then, using such peptides, I should be able to create a vaccine against the hepatitis virus itself, for immunization with such substitutes should create an immune response to both the mimics and the hepatitis virus itself.

Novel vaccines lie to hand. And the advantages of this approach may prove important. First, one does not need to know the pathogen at all! Suppose that during the Gulf War, Saddam Hussein had tossed some virulent bacterial species over the fence, in defiance of global treaties. Blood serum from affected troops would have high concentrations of antibodies against the pathogens. Use such serums to isolate novel peptides binding to those antibodies, and then show that the peptides are not bound by the serum of uninfected, healthy people. Thus the novel peptides must mimic the epitopes on the virulent, but unknown bacterial species. We are on our way to a vaccine against the unknown pathogen. May such work eventually eliminate even the thought of biological warfare.

There are further advantages to the key–lock–key approach. In the current cloning approach to vaccine production, the epitope that elicits the immune response must be a protein. Then the gene encoding that protein is cloned and used to manufacture it. But for many pathogens, the epitopes that trigger an immune response are not protein epitopes, but other kinds of molecules such as complex polymers of sugars called carbohydrates. Many bacteria, for example, have carbohydrate epitopes. Old work in immunology shows that the key–lock–key idea will work even if the first key, the pathogenic epitope, is a carbohydrate molecule. Nevertheless, the second key can be a peptide! In short, peptides can mimic other classes of molecules. Actually, you might have suspected this. The sugar substitute Equal tastes sweet, but it is actually a tiny peptide made of two amino acids. A dipeptide mimics sugar. In fact, peptides appear to be nearly universal shape-mimickers.

The global medical importance of the potential to create vaccines so easily is that it may help us in the fight against orphan diseases. These are diseases that may afflict few or many people, but for which the market is too small to warrant the typical costs of drug development, which is often on the order of $200 million. This cost is an obvious barrier to development for many drugs, including many vaccines. I hope that applied molecular evolution will help bring treatments to many orphan diseases by lowering the cost of drug and vaccine development.

Will all this work? It is beginning to and can only accelerate. In the summer of 1990, three articles appeared almost simultaneously using the same idea. The researchers who wrote the articles realized that rather than merely making libraries of more or less random genes that express random proteins, it would be valuable to create a library of viruses in which each virus expressed a different random peptide on its outer coat. In this way, each virus would carry a molecular tag, its unique peptide. Since the tag would be on the *outside* of the virus, this would make it easy to use molecules binding the tag itself to isolate the virus carrying the tag.

The idea is then to carry out the key–lock–key approach to finding needles in haystacks. Start with the first key, in this case a specific sequence of six amino acids: a hexamer peptide. Then make an antibody to this hexamer. The antibody is the lock. Then use the antibody lock and a large library of random hexamer peptide tags to find second-key hexamers that bind to the antibody lock. The second keys are candidates to mimic the first hexamer key. I should stress that the first experiments were meant to show the practicality of the key–lock–key idea, not to find medically useful molecules.

In the earliest efforts, scientists started with random DNA sequences that encoded strings of six amino acids and cloned them into the gene encoding the outer-coat protein of a specific kind of virus. Thus any such virus would have its specific six-amino-acid sequence "displayed" as a tag on its surface. The number of possible six-amino-acid sequences of 20 kinds of amino acids is 20^6, or about 64 million. Libraries of 21 million of the possible 64 million different peptides were created this way.

With the library prepared, the scientists chose a specific hexapeptide as the first key, and then obtained antibody locks fitting that key. Now the idea was to use the antibody-lock molecules to "fish out" the viruses carrying the second-key hexamer tags. To do this, the antibody molecules were chemically bound to the bottom of a petri plate. The library of 21 million different viruses was incubated with the plate. The hope was that a few of the viruses would have hexamer tags that would be second keys that bound to the antibody-lock molecules. The viruses that did not bind to the antibody molecule were washed off, and then the conditions were altered to release any viruses whose hexamer tags had bound to the antibody locks. Using these procedures, these experiments succeeded in finding second keys.

The results were startling and established a number of critical points: libraries of about 21 million hexapeptides were searched, and 19 second-key peptides were found. So the probability that two hexapeptides look enough alike to bind to the same antibody molecule is about one in a million. Finding molecular mimics is not incredibly hard. The ratio of needles to hay in the haystack is about one needle to a million straws. On average, these hexapeptides differed from the initial hexapeptide in three of the six amino-acid positions. One differed in all six positions! Thus molecules with similar shapes can be scattered far apart in protein space. This was very surprising to most biologists. It is a familiar idea that nearly identical molecules often have nearly identical shapes. These results support a growing suspicion that very different molecules can have very similar shapes! This is a very important fact to which I will return.

The race is now on. Biotechnology companies and drug companies in the United States, Europe, Japan, and elsewhere are starting to invest money in the new approach to drug discovery: don't try to design your molecule. Be smart by being dumb. Make enormous libraries of molecules and learn to select what you want.

The growing passion is well founded. Useful molecules are not limited to proteins and peptides. RNA and DNA molecules may also be discovered and improved through applied molecular evolution. A small company called Gilead is testing a DNA molecule it has evolved that binds to thrombin, a molecule active in forming blood clots. The hope is that this DNA molecule may help prevent harmful blood clotting.

Who would have thought that DNA, the stable carrier of genetic information, could be used to create a molecule that binds thrombin and may prove clinically useful? Nor are DNA molecules able to bind such ligands the only useful DNA molecules we should seek. Remember that the 100,000 genes in each cell in your body turn one another on and off by making proteins that bind to special DNA sites that control the activities of nearby structural genes. Nothing prevents us from generating libraries of roughly random DNA sequences and evolving entirely novel sequences, inserting these into genomes, and thereby controlling cascades of gene activities. Nor are we precluded from evolving novel genes encoding novel proteins that can bind to the cell's own regulatory genes and thereby alter gene activity patterns. For example, cancer arises in many cases because cellular genes called suppressor oncogenes become mutated. Suppressor oncogenes act by turning off cancer genes, called oncogenes, in cells. Nothing should prevent us from evolving entirely novel suppressor oncogenes whose protein products can turn off cellular cancer genes. A pharmacopoeia beyond our imagining lies awaiting.

Jack Szostak and Andy Ellington, at Harvard Medical School, published one of those mind-popping articles a few years ago. They found that they could screen through 10 trillion random RNA molecules simultaneously and fish out RNA molecules that bound to a small organic molecule. Ten trillion at the same time?

You already know that RNA molecules can catalyze reactions. Szostak and Ellington wondered whether they could find RNA molecules that would bind to arbitrary small molecules. If so, such RNA molecules might be useful as drugs or might thereafter be modified to create entirely new ribozymes. They created what biochemists call an affinity column, carrying a specific small organic molecule, a dye. Such columns typically are formed with some kind of beads to which the dye molecules are attached. The idea is that a solution will be poured through the column and flow by the beads. Any molecules in the solution that bind to the dye on the beads will be retained; molecules that do not bind to the dye on the beads will pass swiftly through the column and flow out at its bottom. They created libraries of about 10 trillion random RNA molecules. The library of 10 trillion RNA sequences was poured over each column. Some of the RNA sequences bound to the dye on the column. The rest passed rapidly through the column. The binding RNA sequences were removed by changing the chemical conditions and washing them out. After some further steps, the authors were able to demonstrate that they had "fished out" about 10,000 sequences able to bind selectively to the dye. Since these workers started with 10 trillion sequences, and 10,000 of these bind to the dye, the chances that a randomly chosen RNA sequence binds to one of their dyes is about 10,000 divided by 10 trillion, or one in a billion.

This is stunning. The results suggest this: pick an arbitrary molecular shape, a dye, an epitope on a virus, a molecular groove in a receptor molecule. Make a library with 10 trillion random RNA sequences. One in a billion will bind to your site. One in a billion is a tiny needle in a large haystack. But the staggering thing to realize is that we can now generate and search, in parallel, among trillions of kinds of molecules *simultaneously* to fish out the molecules we want, the candidate drugs we seek.

We saw in Chapter 6 that the estimated protein diversity of the entire planet might be about 10 trillion. Szostak and Ellington, and now many others, are almost trivially generating this diversity of RNA sequences in tiny test tubes and simultaneously screening all of them for the tiny faction that has a desired function. We now rival the planet's diversity in a tiny test tube. And where evolution may have taken eons to fish out its choices, we can already fish out ours in hours.

Prometheus, what did you start?

Universal Molecular Toolboxes

We have now seen that peptides can mimic other molecules. We have seen that there is a one-in-a-million chance that a peptide will bind to an antibody molecule and a one-in-a-billion chance that an RNA molecule will bind to a small organic dye. This is startling enough, but the facts we have discussed portend an even greater shift in how we think about molecular function. It becomes reasonable to consider the hypothesis that finite collections of polymers, proteins, DNA, RNA, or other molecules may be *universal toolboxes* able to carry out essentially any desired function. The diversity required to have a full toolbox may be on the order of 100 million to 100 billion polymers.

The idea of universal molecular toolboxes, I warn you, is heretical. If it is correct, however, the implication is that we can use them to create almost any molecular function we want.

The central idea is simple: there are many more possible sequences than there are shapes! Among proteins with 100 amino acids, there are, as we have seen, about 10^{120} sequences. But how many effectively different shapes can these form? No one knows. Good arguments suggest that only about 100 million effectively different molecular shapes exist on the atomic-size scale where molecules are able to act on one another.

My friends Alan Perelson at Los Alamos and George Oster at Berkeley are the first researchers I am aware of to state this problem crisply. Perelson and Oster were wondering why the human immune system seems to have a diversity of about 100 million antibody molecules. Why not trillions? Why not 42, the universal answer to everything in *The Hitchhiker's Guide to the Galaxy?* Perelson and Oster define an abstract "shape space" with three dimensions for the three spatial dimensions, and then some other dimensions for features of molecules, such as net electric charge, how soluble in water a bump on the molecule is, and so forth. Think of shape space as a kind of N-dimensional box or room. Any shape is a point in this room. Now here is the first crucial point. When an antibody molecule binds an antigen, the fit is not perfect, only approximate, just like a skeleton key fits well enough but not precisely. So one can think of an antibody molecule as binding a "ball" of similarly shaped antigens in shape space.

Within the concept of a shape space, there is an obvious idea and two far deeper ideas. The obvious idea is that similar molecules can have similar shapes, and hence lie in the same ball in shape space. The first deeper idea is this: Since each antibody covers a ball in shape space, a finite number of balls will cover all of shape space! This is very important. Since molecular recognition is sloppy, not precise, a finite number of antibody molecules can fit all possible shapes! A finite number of antibody molecules is a universal toolbox to recognize all possible shapes. A finite number of molecular skeleton keys suffices to fit all molecular locks.

How many keys will suffice? How many balls are required to cover shape space? To answer this, we need to know how big one ball is.

To estimate the volume of one ball in shape space covered by one antibody molecule, Perelson and Oster make use of the fact that the simplest immune systems, such as those in newts, have a diversity of about 10,000 antibody molecules. Now to be useful in evolution, such a repertoire of antibody shapes had best cover a reasonable fraction of shape space, or else it would be of no selective advantage. Only by binding a significant portion of shape space would the immune system be able to bind to a significant number of invaders. Perelson and Oster guess that the minimal useful fraction to bind is $1/e$, or 37 percent, of shape space. They then can calculate that each of the 10,000 antibody molecules covers 1/10,000 of 37 percent of shape space. So they have an estimate of the volume of one ball in shape space. Humans have a repertoire of about 100 million different antibody molecules. Since antibody molecules have more or less random binding sites, covering more or less randomly located balls in shape space, Perelson and Oster then ask what would happen if 100 million balls of this volume were randomly tossed into shape space. Picture Ping-Pong balls that can interpenetrate, tossed at random into your living room. How many would be needed to pretty much fill up the entire room, except for a woeful begonia? Perelson and Oster use mathematical arguments about locations of random Ping-Pong balls to show that 100 million would saturate, or cover, almost all of shape space.

So 100 million skeleton keys are all it takes. There are only about 100 million effectively different shapes, even though there are hyperastronomical numbers of possible polymers and other molecules with these shapes.

Here is the second deep idea about shape space: not only do similar molecules have similar shapes, but small parts of very different molecules, a few tens of atoms each, can have the "same" local shape. Chemically very different molecules, such as peptide and carbohydrate epitopes, or endorphin and opium, can have molecular features that locally look the same, and hence lie in the same ball in shape space, even though the atoms involved are not the same.

So we reach the critical conclusion: your immune system, making about 100 million antibody molecules, is probably able to recognize almost any possible antigen! Stated another way, your own personal immune system is already a universal toolbox able to recognize any molecular epitope.

Antibody molecules are not unique in this capacity. Universal toolboxes can almost certainly be constructed by sufficient diversities of many kinds of molecules. One should be able to create universal toolboxes to bind any epitope using random proteins or random RNA molecules. A universal toolbox of about 100 million to 100 billion molecules should suffice to bind essentially any molecule whatsoever.

Binding is one thing. Catalysis is another. Yet we are led to a further amazing possibility: a finite collection of polymers may be able to function as a universal enzymatic toolbox. If so, a library of some 100 million or 100 trillion molecules might suffice for any catalytic task whatsoever.

Perelson and Oster introduced us to shape space. Let us take the step to what we will call catalytic task space. Think of an enzyme as carrying out some catalytic task—binding to the transition state of a reaction—thereby encouraging a reaction that otherwise would not be likely to occur. By binding to the transition state, the enzyme catalyzes the reaction. In shape space, similar molecules can have similar shapes, but very different molecules can also have the same shape. So too, in catalytic task space, similar reactions are similar tasks, but very different reactions can constitute the same task. For example, two chemically different reactions might each have transition states that are very similar to each other. An enzyme binding one transition state would bind the other one as well, and hence the enzyme could catalyze both reactions. Both reactions would lie in the same "ball" in catalytic task space. An antibody covers a ball in shape space, and a finite number of antibody molecules covers the entire volume of shape space. Similarly, an enzyme covers a ball of similar reactions. Hence a finite number of enzymes can cover catalytic task space and be a universal enzymatic toolbox.

Actually, we already know that something like this is very likely to be true; we just have not said it to ourselves. Your personal immune repertoire is probably already a universal enzymatic toolbox.

Recall that it is possible to create catalytic antibodies—that is, antibody molecules able to catalyze a reaction. To do so, one would like to immunize with the transition state of a reaction and obtain an antibody molecule able to bind the transition state, and hence catalyze the reaction. Unfortunately, the transition state is unstable and persists for only nanoseconds, so one cannot immunize with the transition state itself. Instead, it is possible to use some other molecule whose stable form looks like the transition state of the reaction. So the stable mimic is a second shape representing the same catalytic task as does the transition state. One immunizes with the stable analogue of the transition state. And one in 10 of the antibodies raised against the stable analogue of the transition state actually catalyzes the initial reaction.

We noted that we think that the human immune repertoire of about 100 million antibody molecules is shape universal, that one or more antibody molecules can bind to and recognize any antigen. But that implies that the human immune repertoire may also already be a universal enzymatic toolbox. If so, your own immune system harbors antibody molecules that can catalyze virtually any reaction.

Applied molecular evolution promises enormous practical benefits. Using DNA, RNA, and protein libraries, new drugs, vaccines, enzymes, DNA regulatory sites controlling cascades of gene activities in genomic regulatory networks, and other useful biomolecules lie waiting to be discovered. But there is more lying in our future, just beyond Sinai.

Random Chemistry

Chapter 6 discussed the fundamental supracritical character of the biosphere and its probable role in the origin of life. We saw that molecular reaction systems of sufficient diversity could catch fire and create collectively autocatalytic sets. And we saw that such systems might exhibit subcritical behavior or supracritical behavior. In the latter case, as the variety of organic molecules and of candidate enzymes are increased above a threshold, the chemical reaction system explodes in molecular diversity. Similar principles may help our efforts to discover useful new molecules. I call this effort random chemistry.

Peptides and proteins are, at present, of rather limited value as pharmacological agents. The reason is simple. Proteins, like meat, are digested in your gut. Protein drugs cannot be taken orally. Drug companies prefer small organic molecules, which can be taken orally.

Large pharmaceutical companies have invested decades of effort to amass libraries of hundreds of thousands of small organic molecules that are screened for useful medicinal properties. These libraries are either synthesized or derived from sampling bacteria, tropical plants, and so forth. (As you know, part of the argument for preserving genetic diversity lies in its potential medical benefit. I personally find this argument appalling. Sheer respect for the fruits of 4 billion years of evolution ought to be enough to persuade all of us. Have we lost all sense of humility and awe?)

Our investigation of subcritical and supracritical behavior of chemical systems suggests that it may be possible to surpass, on the vastest scale, library diversities of a "mere" 100,000 or so. Rather, we can aim to generate billions or trillions of novel organic molecules and seek a pharmacopoeia in this explosion.

How? Create a supracritical explosion. Recall our experiment using organic molecules and antibody molecules as our candidate enzymes (Figure 6.3). Take about 1,000 different organic molecules, and place them in solution. Now stir in about 100 million kinds of antibody molecules, our candidate enzymes. If this mixture is supracritical, then over time thousands, then millions, then billions of novel kinds of organic molecules will form. Some reaction pathways will flow rapidly; some, very slowly. But some points are obvious: if we begin with 1,000 kinds of molecules at millimolar concentrations, and the diversity explodes a millionfold to create a diversity of a billion types of organic molecules, then at equilibrium the average concentrations will decrease about a millionfold. This means that concentrations, on average, of the high-diversity library will be in the billionth of a molar, or nanomolar, range. That is actually pretty high. Many cell receptors bind their hormone ligands when the hormone is in this concentration range, or even lower. It is just such responses that we need to fish out interesting molecules.

Add to our soup some cell receptor that responds in the nanomolar range. To be simpleminded, let us use a cloned receptor for a hormone such as estrogen. In addition, let us add to the soup a small concentration of very "hot" radioactively labeled estrogen. The labeled estrogen will bind to its receptor tightly. Now if, in the supracritical reaction mixture, one or more novel molecules exist that bind the estrogen receptor with about the same affinity as estrogen, those novel molecules will displace the hot bound estrogen from its receptor. This displacement can be detected by measuring the released radioactive estrogen molecules. Thus we can know that some unknown types of molecules in our supracritical reaction mixture have been formed and mimic estrogen well enough to bind to the estrogen receptor. Such unknown molecules are candidate drugs to mimic or modulate estrogen itself.

Now all we have to do is isolate the unknown molecule and find out what it is. That molecule was presumably synthesized by a sequence of reactions building it up from the founder set of 1,000 organic molecule building blocks we started with. Suppose, to be concrete, four reactions were catalyzed from the founder set, building up seven molecules to create the unknown estrogen mimic.

Then proceed as follows: create 32 vessels. In each, place all 1,000 molecular building blocks. But, in each, place a different randomly chosen 50 percent of the 100 million types of antibody enzyme molecules. Now the chance that any pot has the critical four antibody molecules catalyzing the critical four reactions is $1/2 \times 1/2 \times 1/2 \times 1/2$, or $1/16$. Thus, by chance, two of the 32 pots should have the critical set of four types of antibody molecules. Now run the reactions in the 32 pots. Two of the 32 should make the unknown estrogen mimic, as tested by displacement of labeled estrogen from the estrogen receptor. Pick one of these two pots, each of which contains a random 50 percent of the initial mix of antibodies. Now you have *reduced* the antibody diversity by a factor of 50 percent. Repeat the "halving" process about 26 times, and you are down to the set of about four antibody molecules critical to make the unknown estrogen mimic. In short, you have winnowed down to the set of four catalytic antibodies needed to synthesize the mimic from the 1,000 building blocks, and you have done so without knowing what the mimic is as yet. As you throw away other antibody molecules, the number of side reactions that are catalyzed will diminish. Why? The side reactions were catalyzed by some of these antibody molecules. As you eliminate the antibody molecules, the side reactions are no longer catalyzed; therefore, the concentration of the unknown mimic will increase. In the end, you have, on the one hand, a high enough concentration of the unknown estrogen mimic to characterize it and synthesize it by any means you choose. On the other hand, you have selected a set of catalytic antibodies to synthesize an estrogen mimic without needing to know beforehand what the probable structure of that mimic might be. You have carried out a process that might be called random chemistry.

An unimaginably enormous pharmacopoeia is within reach.

Experimental Creation of Life?

Homo ludens, playful man. I have described two areas of high potential medical significance: applied molecular evolution and random chemistry. Scientific discovery is a mystery in any case, and a mystery of synchronicity. Many people are converging on these ideas from many directions. In my own case, ideas about both applied molecular evolution and random chemistry came from playing. In fact, I was playing with dreams about how life might emerge naturally, a near inevitable expression of complex chemistry. I found my way to autocatalytic sets, about which you now know. That theory is based on models of the probability that a randomly chosen protein catalyzes a randomly chosen reaction. One day, listening to a seminar on the experimental evolution of enzymes to catalyze new reactions, I was struck by an idea: Why not make random proteins and actually find out the probability that a randomly chosen protein catalyzes a randomly chosen reaction?

Why not, indeed? Applied molecular evolution.

And if chemical reaction graphs exhibit subcritical and supracritical behavior, why not mix antibody molecules and organic molecules and catalyze a supracritical explosion?

Why not, indeed? Random chemistry looms.

But if applied molecular evolution now promises a harvest of novel molecules, it also offers a harvest of polymers able to catalyze randomly chosen reactions. And if random chemistry promises libraries of organic diversity to winnow for tomorrow's medicines, it also promises explosions of chemical diversity.

Why not ask if one can create a collectively autocatalytic set?

Why not, indeed? I shall not be overly surprised if in the coming decades, some experimental group creates such life anew, snapping into existence in some real chemostat, creating protocells that coevolve with one another to the subcritical–supracritical boundary. I would not be overly surprised. But I would be thrilled.

Homo ludens needs *Homo habilis* and *Homo sapiens*. We stand on the verge of creating a vaster diversity of molecular forms in one place and time than ever before, we may assume, in the history of the earth, perhaps in the history of the universe. A vast wealth of new useful molecules. An unknown peril of fearful new molecules. Will we do this? Yes, of course we will. We always pursue the technologically feasible. We are, after all, both *Homo ludens* and *Homo habilis*. But can we, *Homo sapiens,* calculate the consequences? No. Never could, never will. Like the grains in the self-organized sandpile, we are carried willy-nilly by our own inventions. We all stand in danger of being swept away by the small and large torrents of change we ourselves unleash.

Chapter 8

High-Country
Adventures

"Wedges in the economy of nature," wrote Darwin in his diary, leaving us with a glimpse of his own first glimpse of natural selection. Each organism, the fitter and the less fit, would wedge itself into the filled nooks and crannies of the tangled bank of life, struggling against all others to jam itself onto the wedge-filled surface of possibilities. Nature, red in tooth and claw, was the nineteenth-century image of natural selection. And natural selection would ever sift out the fitter from the less fit, ceaselessly screening, such that, down the eons, adapted forms would accumulate useful variations and proliferate. Later biologists, by the fourth decade of the twentieth century, would invent the image of an adaptive landscape whose peaks represent the highly fit forms, and see evolution as the struggle of populations of organisms driven by mutation, recombination, and selection, to climb toward those high peaks.

Life is a high-country adventure.

Make no mistake: the central image in biological sciences, since Darwin, is that of natural selection sifting for useful variations among mutations that are random with respect to their prospective effects on the organism. This image fully dominates our current view of life. Chief among the consequences is our conviction that selection is the sole source of order in biology. Without selection, we reason, there could be no order, only chaos. We the unexpected; we the very lucky.

Yet in this book we have seen profound sources of spontaneous order. Is it so clear that without selection there could be no order? I cannot think so. I believe we are entering a new era in which life will be seen as a natural expression of tendencies toward order in a far from equilibrium universe. I believe phenomena ranging from the origin of life to the order of ontogeny to the poised order of ecosystems we will

149

discuss in Chapter 10 are all parts of order for free. Yet I, too, am a Darwinian, persuaded of the efficacy of natural selection. We stand in the deepest need of a new conceptual framework that will allow us to understand an evolutionary process in which self-organization, selection, and historical accident find their natural places with one another. We have no such framework as yet. The aim of this chapter is to show how selection works in its high-country adventure, but also to show its limitations, and then to attempt to sketch part of a new marriage of self-organization, selection, and accident.

It is one thing for me to tell those of you who are not biologists that self-organization plays a role in this history of life. Why not, you might say. And I agree that it seems obvious. After all, lipids in water do form hollow bilipid membrane spheres, such as cell membranes, without the benefit of natural selection. Selection need not accomplish everything. But nonbiologists typically do not understand the profound power the selection-is-the-only-source-of-order view holds for post-Darwinian biologists. Prior to Darwin, the Rational Morphologists, confronting what they believed were fixed species, could seek laws of form in the morphologies they collected. The similarities of the vertebrate limb from reptiles to birds to mammals is a familiar example. But for all their attempts, those eighteenth- and early-nineteenth-Century biologists could find no ready explanation for the order exhibited by organisms. With Darwin, a brand new, radically new, idea was born. While Darwin's compatriot Bishop William Paley, could argue that the order of a watch might presuppose a watchmaker, and hence the order of organisms required a divine watchmaker, in Darwin's theory the order of organisms bespoke no God, but the novel mechanism of natural selection ceaselessly sifting. So biologists now tend to believe profoundly that natural selection is the invisible hand that crafts well-wrought forms. It may be an overstatement to claim that biologists view selection as the sole source of order in biology, but not by much. If current biology has a central canon, you have now heard it.

The canon of selection as the sole or even primary source of order is the quintessential statement of we the accidental—we who could easily be different in almost all ways, or not here at all.

Again, make no mistake, this is the central, settled view of almost all contemporary biologists. While I am spending quite a bit of my scientific life attempting to explore reasons why this view is deeply inadequate, it is important to admit that the standard view has much to commend it. Biologists see organisms as tinkered-together contraptions, and evolution as a tinkerer. Organisms are Rube Goldberg machines; the jawbone of an early fish became the inner ear of a mammal. Organisms really are full of the strangest solutions to design problems. Biologists delight in discovering these and noting to one another, and particularly to those of us inclined toward theory in biology, "You'd never have predicted that!" Inevitably, the assertion is correct. Organisms do find the strangest ways of doing things. The full-blooded awareness of this must always be borne in mind. And you, reading the ideas of a biologist who holds renegade views, must be forewarned. Renegade views are not correct merely because they are renegade, written in a book, and even interesting. But I hope to persuade you that there are compelling reasons to believe they may prove true.

It is easy to find evidence of apparent ad hocery. Pick up a pinecone and count the spiral rows of scales. You may find eight spirals winding up to the left and 13 spirals winding up to the right, or 13 left and 21 right spirals, or other pairs of numbers. The striking fact is that these pairs of numbers are adjacent numbers in the famous Fibonacci series: 1, 1, 2, 3, 5, 8, 13, 21 . . . Here, each term is the sum of the previous two terms. The phenomenon is well known and called phyllotaxis. Many are the efforts of biologists to understand why pinecones, sunflowers, and many other plants exhibit this remarkable pattern. Organisms do the strangest things, but all these odd things need not reflect selection or historical accident. Some of the best efforts to understand phyllotaxis appeal to a form of self-organization. Paul Green, at Stanford, has argued persuasively that the Fibonacci series is just what one would expect as the simplest self-repeating pattern that can be generated by the particular growth processes in the growing tips of the tissues that form sunflowers, pinecones, and so forth. Like a snowflake and its sixfold symmetry, the pinecone and its phyllotaxis may be part of order for free.

Important premises are presupposed in the standard canon, according to which organisms are ad hoc contraptions. The most important presupposition—and, indeed, one of the most important presuppositions of Darwin's entire thesis—is *gradualism,* the idea that mutations to the genome, or genotype, can cause *minor* variations in the organism's properties—that is, in its phenotype. Furthermore, the bric-a-brac view assumes that these minor useful variations can be *accumulated* piecemeal, bit by bit, over the eons to create the complex order found in the organisms we observe.

But are these claims true? Is it obvious that "gradualism" always works? And even if the claims of gradualism are true for contemporary organisms, need they be true? That is, can all complex systems be "improved" and ultimately assembled by accumulating a succession of minor modifications? And if it is not true of all complex systems, but is true of organisms, then what are the properties that permit some complex systems to be assembled by an evolutionary process? Further, what is the source of these properties, this ability to evolve? Is evolution powerful enough to *construct* organisms that are able to adapt by mutation, recombination, and selection? Or is another source of order—spontaneous self-organization—required?

It is fair to say that Darwin simply assumed that gradual improvement was possible in general. He based his argument on the selection carried out by breeders of cattle, pigeons, dogs, and other domesticated plants and animals. But it is a long, long step from selection by hand for alteration in ear shape to the conclusion that all features of complex organisms can evolve by the gradual accumulation of useful variations.

Darwin's assumption, I will try to show, was almost certainly wrong. It does not appear to be the case that gradualism always holds. In some complex systems, any minor change causes catastrophic changes in the behavior of the system. In these cases, as we will soon discuss, selection cannot assemble complex systems. Here is one fundamental limit to selection. There is a second fundamental limit as well. Even when gradualism does hold in the sense that minor mutations cause minor changes in phenotype, it still does not follow that selection can successfully accumulate the minor improvements. Instead, an "error catastrophe" can occur. An adapting population then accumulates a succession of minor catastrophes rather than a succession of minor improvements. Even with selection sifting, the order of the organism melts silently away. We will discuss error catastrophe later in the chapter.

Selection, in short, is powerful but not all-powerful. Darwin might have realized this were he familiar with our present-day computers.

Let us imagine trying to evolve a computer program that calculates something modestly complicated, such as the trajectories of three mutually gravitating objects or the seventh root of an arbitrary real number. Start with an arbitrarily chosen computer program written, as is always possible, as a string of 1s and 0s, and then randomly flip bits, changing 1s to 0s and vice versa. Then test each mutated program with a set of input data to see if it carries out the desired computation and selects fitter variants. If you try this task, you will find it is not so easy. Why not? And what would happen if, in addition to attempting to evolve such a computer program, we were more ambitious and attempted to evolve the shortest possible program that will carry out the task? Such a "shortest program" is one that is maximally compressed; that is, all redundancies have been squeezed out of it.

Evolving a serial computer program is either very hard or essentially impossible because it is incredibly fragile. Serial computer programs contain instructions such as "compare two numbers and do such and such depending on which is larger" or "repeat the following action 1,000 times." The computation performed is extremely sensitive to the order in which actions are carried out, the precise details of the logic, numbers of iterations, and so forth. The result is that almost any random change in a computer program produces "garbage." Familiar computer programs are precisely the kind of complex systems that do not have the property that small changes in structure yield small changes in behavior. Almost all small changes in structure lead to catastrophic changes in behavior. Furthermore, this problem becomes worse as redundancy is squeezed out of the program in order to achieve a minimal program to perform the algorithm. In a nutshell, the more "compressed" the program, the more catastrophically it is altered by any minor change in the instructions. Hence the more compressed the program, the harder it is to achieve by any evolutionary search process.

And yet the world abounds with complex systems that have successfully evolved—organisms, economies, our legal system. We should begin to ask, "What kinds of complex systems can be assembled by an evolutionary process?" I should stress that no general answer is known, but that systems with some kinds of redundancy are almost certainly far more readily evolved than those without redundancy. Unfortunately, we only roughly understand what "redundancy" actually means in evolving systems.

Computer scientists have defined a concept called algorithmic complexity. A computer program is a finite set of instructions that, if followed, computes the desired result. In an intuitive sense, "hard" problems might be defined as those requiring more instructions than "simple" problems. The algorithmic complexity of a program is defined as the length of the shortest program that will carry out the computation. In actual practice, one cannot, in general, prove that a program is the minimal one. A shorter one might always exist. However, quite clever uses have been made of this measure. In particular, the following can be shown: a minimal program is one for which the symbol string has no internal redundancy. For example, suppose that the true minimal program were (1010001 . . .), but one had in hand a program with each "bit" duplicated: (11001100000011 . . .). In this example, redundancy could obviously be eliminated by replacing each double bit by one bit. Or there may be more subtle patterns that can be compressed. Perhaps after eliminating the double bits we find that with every 226th bit, an identical pattern of 1s and 0s occurs. With a little recoding, this repeating pattern can be replaced by a routine that says, every 226 bits insert XYZ. And so the program becomes shorter still. If we could detect and remove all the redundancies, the result would be a minimal program that is maximally compressed. It would be patternless—no more redundancies could be squeezed out of it. It follows that any such minimal program cannot be distinguished from a random sequence of 1 and 0 symbols! If we could find a pattern, there would be a redundancy to remove.

It seems likely that there is no way to evolve a maximally compressed program in less time than it would take to exhaustively generate all possible programs, testing each to see if it carries out the desired task. When all redundancy has been squeezed from a program, virtually any change in any symbol would be expected to cause catastrophic variation in the behavior of the algorithm. Thus nearby variants in the program compute very different algorithms.

Adaptation is usually thought of as a process of "hill climbing" through minor variations toward "peaks" of high fitness on a fitness landscape. And natural selection is thought of as "pulling" an adapting population toward such peaks. We can imagine a mountain range on which populations of organisms (or in this case, programs) are feeling their way to the summits. Depending on whether it is beneficial, a random change in the genome (the computer code) puts a mutant higher or lower on the terrain. If the mountain terrain is rugged, but looks like familiar mountains, the terrain is still smooth enough to provide clues in the immediate vicinity about which directions to take. There are pathways uphill to the distant peaks, and natural selection, in sifting for the fitter variants, pulls the population up toward them.

The search problem is compounded by the fact that the evolving population cannot actually see the contours of the landscape. There is no way to soar above and take a God's-eye view. We can think of the population as sending out "feelers" by generating, at random, various mutations. If a mutation occupies a position higher on the terrain, it is fitter, and the population is pulled to the new position. Then random mutations from that position feel in all directions. Selection again pulls the population one step farther uphill. Here is gradualism at work in Darwin's sense. Undoubtedly, organisms do evolve by such gradual ascent. Undoubtedly, people also often solve complex design problems piecemeal, assembling good designs by a succession of trial-and-error searching. The process of evolving organisms and artifacts is deeply similar—a theme I will devote Chapter 9 to exploring.

But when all minor variations cause catastrophic variations in the behavior of the system, organism, or artifact, the fitness landscape is essentially random; hence no local cues exist to detect directions of change that are uphill toward the distant peaks. We will see this in detail later, but for the moment imagine yourself on an utterly jagged moonscape, perched on a ledge with cliffs plunging downward in some directions and vaulting upward to various heights in others. Nowhere can you see a long distance away. Although there may be very high pinnacles nearby, you cannot know this. There are no local clues to tell you which way to go. Search becomes mere random search.

But once search is merely random, with no clues about uphill trends, the only way to find the highest pinnacle is to search the whole space! Rather than seeing Mont Blanc from Chamonix and heading uphill for it, wine, cheese, paté, and bread in your pack—or better, real climbing gear—you have to explore all of the Alps methodically, trying out each square meter, to find the peak of Mont Blanc.

Maximally compressed computer programs almost certainly cannot be evolved in less time than it takes to search through the entire space of possible programs. Such systems are not the kind of complex system that can be evolved by natural selection in the lifetime of the universe. Suppose the minimal program we are searching for required N bits. Then the space of all possible programs of that length is 2^N. For a relatively small program of 1,000 bits, $N = 1,000$, we get the familiar kind of hyperastronomical number: $2^{1,000}$, or 10^{300}. Now the first conclusion is this: because the program is maximally compressed, any change will cause catastrophic alterations in the computation performed. The fitness landscape is entirely random. The next fact is this: the landscape has only a few peaks that actually perform the desired algorithm. In fact, it has recently been shown by the mathematician Gregory Chaitin that for most problems there is only one or, at most, a few such minimal programs. It is intuitively clear that if the landscape is random, providing no clues about good directions to search, then at best the search must be a random or systematic search of all the 10^{300} possible programs to find the needle in the haystack, the possibly unique minimal program. This is just like finding Mont Blanc by searching every square meter of the Alps; the search time is, at best, proportional to the size of the program space.

So we reach our conclusion. If there are 10^{300} possible programs to search through, and one must try them all to be sure of finding the best one, and one could "try" a different program every billionth of a second, it would take 10^{291} seconds to find the best program. Here we are again, trying to grasp numbers spanning across time scales impossibly larger than the history of the universe.

You might be thinking of a counterargument. "Fine," you might say, "evolution cannot assemble a maximally compressed program, or maximally compressed organism, whatever that might be, right off the bat.

But maybe evolution can, in fact, assemble such a maximally compressed program, or organism, by first evolving a redundant program or organism, and then squeezing it down to maximal compression." You might, in short, concede that evolution could not find its way to the minimal program, length 1,000, if evolution were confined from the start to search only the space of minimal programs, in less time than the history of the universe. But what if the search process began with a highly redundant program and gradually compressed it to the minimal length, $N = 1,000$? Could such a procedure succeed in finding the minimal program? No one knows, but I bet not. Here is the intuition. Evolving a highly redundant program should be easy. If we do not care how long our programs are, there are many, many programs that will carry out the same task. And these highly redundant programs are not as brittle as compressed programs; minor alterations in the code can cause minor alterations in the program's behavior. For example, suppose that one subroutine accomplishes a task. Inserting a duplicate copy of the routine need not alter behavior. Thereafter, mutational alteration of one copy need not destroy function, because the other, redundant copy can still function in its place. Meanwhile, the mutant copy can "explore" new possibilities. Thus redundant algorithms can evolve smoothly. Suppose that a first good program were found having a length two times the minimal length, N, hence $2N$. Then consider trying to slowly squeeze out the redundancy by evolving shorter programs. Starting with the $2N$ program, delete a randomly chosen bit and then flip some bits at random, trying to evolve a good program one bit shorter, length $2N - 1$. Then mutate that shorter program to find a still shorter program, length $2N - 2$, and so on until we reach a minimal program length N.

Here, I believe, is the fundamental flaw: for this process—creeping down toward the minimal length program—to be helpful, it must be the case that the program found at each stage helps locate the good program at the next stage, when the program length decreases by 1. But as we go down the rungs, evolving shorter and shorter programs, each is less redundant—more random—than the one above it. Thus it will be more brittle, less amenable to evolving by random mutation. If so, then at each stage, the program just found provides fewer and fewer clues about where to search for the next shorter program. And I bet that as the minimal length program is approached, the previous program, length $N + 1$, provides no clues whatever about where to search the space of possible programs length N.

I do not know that this idea is correct, although it seems the kind of thing good mathematicians might prove. I might therefore promote the idea to what mathematicians call a conjecture: as the minimum program is approached, search for it would become search on a fully random landscape for which no earlier clues from more redundant programs would be helpful. If so, minimal programs are not the kinds of complex things that can be assembled by an evolutionary process.

The intuition to take away from this example is that not all complex systems can be achieved by adaptive search in reasonable time. Some conditions, largely unknown but involving redundancies, must exist that characterize the kinds of complex systems that can be assembled by an evolutionary search process. For example, John Koza, at Stanford, is actually evolving computer programs to carry out a variety of modestly complex tasks. John finds that his programs are just the kind of bric-a-brac, ad hoc assemblages that biologists tend to see in organisms—they are definitely not highly compressed, but highly redundant in a variety of ways that are still poorly understood. We will return to this after discussing the structure of fitness landscapes and their bearing on adaptive search.

The question of what kinds of complex systems can be assembled by an evolutionary search process not only is important for understanding biology, but may be of practical importance in understanding technological and cultural evolution as well. The sensitivity of our most complex artifacts to catastrophic failure from tiny causes—for example, the *Challenger* disaster, the failed *Mars Observer* mission, and power-grid failures affecting large regions—suggests that we are now butting our heads against a problem that life has nuzzled for enormously longer periods: how to produce complex systems that do not teeter on the brink of collapse. Perhaps general principles governing search in vast spaces of possibilities cover all these diverse evolutionary processes, and will help us design—or even evolve—more robust systems.

Life on Landscapes

Back to the beginning, to the first eons. For almost 3 billion years, life pulsed quiet, still, waiting. For 3 billion years, bacteria perfected their molecular wisdom, flourished in trillions of pools, cracks, hot vents, and crannies, blossoming across the globe in, I believe, a supracritical explosion of molecular forms. The early life-forms linked their metabolisms, traded their molecular products, toxins, nutrients, and simple waste, to create a spreading supracritical wave of molecular diversity. Over those 3 billion years, small, magnificent, stage-setting changes occurred. The nucleated eukaryotic cell arose, harboring—according to one current theory—captured bacteria that became organelles: chloroplasts and mitochondria driving photosynthesis and energy metabolism. The earliest evidence of an organelle bearing nucleated cell is, according to Bruce Runnegar of UCLA, to be found in rocks 2.15 billion years old. These spiral organisms, with complex banding patterns, range up to half a meter in length (Figure 8.1). Runnegar suggests that these organisms may be large single-celled creatures somewhat like modern *Acetabularium* (Figure 8.2). Referring to the complex morphology of *Acetabularia,* with its exquisite umbrella cap on a tall stalk, and other single-celled organisms with detailed morphologies and complex life cycles, embryologist and theoretical biologist Lewis Wolpert says, "The modern eukaryotic cell knows all there is to know. After the eukaryotic cell, it's all downhill." Lewis's point is that almost everything needed to form a multicelled organism is already contained in the eukaryotic cell.

Cells, you see, had to learn to form multicelled organisms; in addition, cells had to learn to form individuals. Not so easy, this evolution of individuals, because most cells in the body die when the organism dies, hence giving up their chance for immortality, dividing forever like bacteria. The cells that die form the soma, or body. Only germ cells, giving rise to sperm or egg, have a chance at eternity. The puzzle of the origin of multicelled organisms, and hence of the individual, is why a colony of genetically identical cells learned to form a multicelled organism. This is our first instance of a general and difficult problem, the "benefits of the individual" versus the "benefit of the group." Leo Buss of Yale University has focused attention on this problem, for the origin of the individual is problematic. Natural selection favors fitter individuals. If the world consists of dividing single-celled eukaryotic organisms, and I am one of them, why is it to my benefit to form part of a multicelled organism where my fate is to die? If only the germ cells have progeny down the eons and the rest of the cells in the multicelled organism are destined to dust, why should those doomed cells be content with their lot? In the case of the evolution of the individual multicelled organism, the answer is modestly clear. Since all cells in the organism have the same genes, all cells are genetically identical. Even if most cells in the organism die, and only the germ cells have progeny, it may be beneficial to suffer the loss in dying individual cells because the multicelled organism itself is more successful than single cells alone in spreading the genetic heritage. Presumably, membership in a multicelled critter affords an enhanced chance to invade new niches and leave lots of offspring.

Multicelled organisms are actually known well before the famous Cambrian explosion 550 million years ago. About 700 million to 560 million years ago, and forming the final Precambrian period, is the Ven-

Figure 8.1 *Ancient ancestor. The earliest known fossil eukaryotic cell,* Grypania spiralis, *a photosynthetic alga, 2.15 billion years old.*

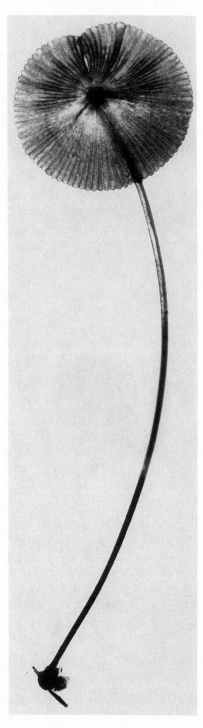

Figure 8.2 Acetabularium, *a single-celled organism with a highly complex mor-phology and life cycle that may resemble early eukaryotic life-forms. Mature height is approximately 1 millimeter.*

dian and the Ediacrian fauna, named for a region in Australia. Here Runnegar and his colleagues are finding wormlike forms, soft-bodied and up to a meter in length, coiled in these ancient rocks (Figure 8.3). Then comes the Cambrian blast of diversity and the rest is, from the perspective of 4 billion years, recent history.

For the 3.45 billion years of life, simple and complex organisms have been adapting, accumulating useful variations, climbing up fitness landscapes toward peaks of high fitness. Yet we hardly know what such fitness landscapes look like or how successful an evolutionary search process is as a function of landscape structure. Landscapes can be smooth and single peaked, rugged and multipeaked, or entirely random. Evolution searches such landscapes using mutation, recombination, and selection. But as we have already seen for maximally compressed programs, such search processes may fail to find the high peaks. Or on smooth landscapes, even when high peaks are found, mutations accumulating in the adapting population may cause the error catastrophe alluded to earlier. The population can melt off the peaks and flow out over the landscape, losing any accumulated useful variants. Clearly, the notion of organisms inexorably becoming fitter through natural selection is not as simple as it seems on the surface. To further appreciate the issues and their implications, it helps to review some of the history of the field known as population biology.

Darwin taught us that organisms evolve by natural selection acting on heritable variations. But Darwin did not know the source of the muta-

Figure 8.3 *An early multicelled wormlike organism,* Dickinsonia, *approximately 1 meter in length and about 700 million years old.*

tions; what it was that was actually changing was obscure for the next half-century. Darwin and others thought in terms of "blending inheritance." Children resemble both parents. However heredity might work, it was supposed that this resemblance to both parents was because of a kind of mixing and averaging, yellow and blue mixing to green. But blending posed a problem: after many generations, the genetic variation in a mating population would dwindle away. If yellow and blue are eye colors, after many generations of blending inheritance the entire population would be green-eyed. There would be no further heritable variation on which selection might act.

Mendel's genetic studies ultimately provided a way out of the problem of blending inheritance, although nearly half a century separated his work and the use made of it, first by the famous mathematician George H. Hardy and the biologist W. Weinberg, and then by Ronald A. Fisher, a young mathematician at Cambridge University. Recall Mendel's peas, and the underlying concept of atoms of heredity. Mendel studied seven well-chosen pairs of traits—rough versus smooth seeds, for example. He found that the first-generation offspring of rough and smooth parents resembled only one parent, but that both forms, or phenotypes, reappeared in the grandchildren. Evidently, the "rough" and the "smooth" atoms of heredity, now called genes, passed unmodified, and unblended, through the children into the grandchildren.

By the early decades of this century, it had become clear that the chromosomes in cell nuclei must be the carriers of genetic information. Further, it was clear that many different genes, affecting many different traits, were located in a line on each chromosome, and that many genes, like "rough" and "smooth," came in such alternative versions, now called alleles. Based on this, Hardy and Weinberg showed that blending of different alleles does not occur in a mating population. Instead, the different alleles, called the genetic variance, will persist in the mating population for selection to act upon.

Mendel's laws set the stage for Fisher to pose a single fundamental question that had to receive an affirmative answer for Darwin's theory to be correct: suppose a single gene in a population has two alleles, say A and a—corresponding to blue eyes and brown eyes—and that the A allele confers a slight selective advantage on its bearer. Can natural selection acting on a population actually increase the frequency of the A allele in the population?

The answer, established by Fisher, J. B. S. Haldane, and Sewall Wright, can be yes. Darwinism can work. Setting the conditions under which this is true laid the foundations of modern population genetics. But the answer, it turns out, depends on the structure of the fitness landscape explored by the adapting population, whether it is smooth and single peaked, rugged and multipeaked, or completely random. We still know very little about the structure of such landscapes, and the efficacy of adaptive search on them. When can mutation, recombination, and selection actually climb the high peaks? In short, almost 140 years after Darwin's seminal book, we do not understand the powers and limitations of natural selection, we do not know what kinds of complex systems can be assembled by an evolutionary process, and we do not even begin to understand how selection and self-organization work together to create the splendor of a summer afternoon in an Alpine meadow flooded with flowers, insects, worms, soil, other animals, and humans, making our worlds together.

Genotype Spaces and Fitness Landscapes

It is time to look in detail at simple models of fitness landscapes. While these models are mere beginnings, they already yield clues about the power and limits of natural selection.

Humans, like almost all higher plants and animals, are diploids. We receive two copies of each chromosome: one from our mother, one from our father. Bacteria are haploids, having only a single copy of their chromosome. Typically, bacteria simply divide; they do not have parents or sex. (This is not quite true. Bacteria sometimes mate and exchange genetic material, but we can safely ignore such bacterial sex in the following discussion.) While most population genetics has developed concerning diploid mating populations, we will focus on haploid populations, such as bacteria, and ignore mating. This restriction helps us gain an image of what fitness landscapes look like. Later, we can add back the complications of diploidy and mating.

To be concrete, suppose we consider a haploid organism with N different genes, each of which comes in two versions, or alleles, 1 and 0. With N genes, each with two alleles, the number of possible genotypes is now familiar: 2^N. The bacterium *E. coli*, living merrily in the intestines, has about 3,000 genes; hence its genotype space might have $2^{3,000}$ or 10^{900} possible genotypes. Even PPLO, pleuromona, the simplest free-living organisms with only 500 to 800 genes, has between 10^{150} and 10^{240} potential genotypes. Now consider diploid organisms, with two copies of each gene. Plants may have 20,000 genes. If each has two alleles, the number of diploid genotypes is $2^{20,000}$ for the maternal chromosomes times $2^{20,000}$ for the paternal chromosomes, or about $10^{12,000}$. Clearly, genotype spaces are vast; even for a modestly long genome, there are an astronomical number of states it can assume. Any population of a species represents at any time a very, very small fraction of the space of its possible genotypes.

Natural selection acts to pick fitter variants. How does this actually happen? Consider a population of bacteria of the same species, one subset of which contains a gene, B, which causes the bacterium to be blue, while the remaining subset contains a different allele of the same gene, R, which causes the bacterium to be red. Suppose that in a specific environment, it is advantageous to be red rather than blue. Then when biologists say the red allele is fitter than the blue allele, we mean, in a rough sense, that the red bacteria would be more likely to divide and leave offspring than the blue bacteria. In the simplest sense, the red bacteria would divide or reproduce, more rapidly. Suppose that the dividing time for the red bacteria is 10 minutes, while the dividing time for the blue bacteria is 20 minutes. Then in an hour, any initial red bacterium would have undergone six doublings, creating 2^6, or 64, red progeny, while any blue bacterium would have undergone three doublings, creating 2^3, or 8, progeny. Both the red and the blue populations will undergo an exponential growth in population as time increases, but the exponential rate of increase is greater for the red than for the blue bacteria. Thus whatever the initial ratio of red and blue bacteria in the population, as long as there is at least one red bacterium, in time the total population will become almost entirely red bacteria, with a small and ever-dwindling fraction of blue bacteria. Selection in favor of the fitter red allele will have increased the frequency of the R gene compared with the B gene in the bacterial population. That is the core of natural selection substituting a fitter for a less fit allele.

I have slipped over a critical detail. Suppose that in the initial population, there were only a single red bacterium and many blue bacteria. Then before the lone red bacterium had a chance to divide, it might happen to be killed. Bad luck. The point here is that such fluctuations and accidental perturbations can, and do, affect the evolutionary process. Roughly speaking, once there is some threshold number of red bacteria, the chance that all of them accidentally make their ways to bacterial nirvana before the red population begins to explode becomes vanishingly small. Once beyond the threshold, fluctuations become very unlikely to stop the remaining inexorable march toward a red bacterial population.

I shall speak later of an adapting population of organisms moving, or flowing, across a genotype space under the drive of natural selection. You should retain the images of the bacteria in mind. If the population of bacteria began as entirely blue, and remained entirely blue, nothing would happen. But suppose a B gene in one bacterium happened to mutate to the R allele. Then that happy red bug and its progeny would divide faster and ultimately displace the blue bacteria. The population would "move" from the genotype with the *B* allele to the genotype with the R allele. (It is true that a tiny dwindling fraction of blue bacteria would remain. However, we can modify our experiment and carry it out in a chemostat, which maintains a constant number of bacteria internally, while food is added, and bacteria plus unused food and waste are removed. Eventually, all the blue bacteria will be diluted out of this finite population system, leaving only red bacteria.)

Now that we have introduced the ideas of a fitness landscape and a genotype space, we want to use these ideas as tools to examine just how effective natural selection can actually be in shaping the organisms of the biosphere. What, then, do these fitness landscapes actually look like? And how does the character of such a landscape govern the successes or failures of an evolutionary assembly process?

Suppose it is possible to measure the "fitness" of each genotype. Let this fitness be thought of as a "height." Fitter genotypes are higher than less fit genotypes. We need one other critical notion, that of "neighboring" genotypes. Consider a genotype with only four genes, each having two alleles: 1 and 0. Thus there are 16 possible genotypes—(0000), (0001), (0010) . . . all the way to (1111). Each genotype is "next to" those that differ by changing any one of the four genes to the other allele. Thus (0000) is next to (0001), (0010), (0100), and (1000). Figure 8.4*a* shows the 16 possible genotypes, each as a corner, or "vertex," of

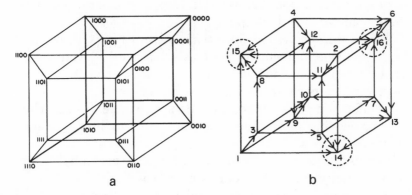

a   b

Figure 8.4 *A Boolean hypercube.* (a) *All possible genotypes of a genome with four genes, each of which can be in two states, or alleles, can be plotted as vertices on a four-dimensional Boolean hypercube. Each genotype is connected to its four neighbors, whose genomes differ by a single mutation. Hence 0010 is "next to" 0000, 0011, 1010, and 0110.* (b) *Each genotype has been assigned, at random, a rank-order fitness, ranging from the worst, 1, to the best, 16. The circled vertices indicate local optima—genotypes that are fitter than all their one-mutant neighbors. Arrows from each genotype point to fitter neighbors.*

what a mathematician would call a four-dimensional Boolean hypercube. The dimensionality of the Boolean hypercube is just equal to the number of neighbors of each genotype. If there are N genes, the number of neighboring genotypes to each genotype is just N, and there are a total of 2^N possible genotypes, each at a different corner of the hypercube.

Consider the following peculiar case: let the 16 genotypes each have a fitness that is "assigned" totally at random—say, choosing at random decimal numbers between 0.0 and 1.0. Now rank the genotypes from least fit, 1, to most fit, 16. Thus to carry out this particular whimsy, it suffices to assign the numbers 1, 2, 3 . . . 16 at random, one to each corner of the Boolean hypercube (Figure 8.4b).

We have just created a *random fitness landscape*. Our decision to assign fitnesses at random means that our random landscape is deeply similar to the fitness landscapes of maximally compressed computer programs. In both cases, changing a single "bit," making any mutation at all, completely randomizes "fitness."

Once we understand the nature of these random landscapes and evolution on them, we will better appreciate what it is about organisms that is different, how their landscapes are nonrandom, and how that nonrandomness is critical to the evolutionary assembly of complex organisms. We will find reasons to believe that it is not natural selection alone that shapes the biosphere. Evolution requires landscapes that are not random. The deepest source of such landscapes may be the kind of principles of self-organization that we seek. Here is one part of the marriage of self-organization and selection.

Consider the simplest version of an *adaptive walk* on a fitness landscape. The rule of our simplified game is this: start at a vertex, or genotype. Consider a one-mutant neighbor—that is, changing a randomly chosen gene to its alternative allele. If the variant is fitter, go there. If the one-mutant neighbor is not fitter, do not go there. Instead, choose at random a one-mutant neighbor again, and move to this new genotype if it is fitter.

This prescription allows us to use arrows pointing "uphill" from each genotype toward those of its one-mutant neighbors (if any), which are fitter (Figure 8.4b). An adapting population, if "released" at one vertex, hence all having the same genotype, will remain there. If, however, a mutation occurs in one of the four genes and the new mutant genotype is fitter, the fitter population will outgrow the less fit population; hence the total population will "shift," or flow, from the initial corner of the hypercube to the corresponding neighboring corner.

So if we think of the fitness of a genotype as a height, the adaptive walk starts at a genotype and climbs uphill. Now we can ask what such adaptive walks on random landscapes look like and why evolution can be difficult.

The first essential feature of adaptive walks is that they proceed uphill until a local peak is reached. Like a hilltop in a mountainous area, such a local peak is higher than any point in its immediate vicinity, but may be far lower than the highest peak, the global optimum. Adaptive walks stop on such local peaks. There is no higher point in the immediate vicinity, so they are trapped, with no way to get to the distant high summits. In the landscape in Figure 8.4b, there are three local peaks. Here we consider tiny landscapes with only four genes and 16 genotypes. What happens in very large genotype spaces? How many local peaks exist?

Random landscapes are chock-full of local peaks. The number of local peaks is a staggering $(2^N)/(N + 1)$. (Actually, it's rather easy to see why this is true. Any genotype has N neighbors. The chance that it is the fittest among its N neighbors and itself, hence a local peak, is $1/[N + 1]$. Since there are 2^N genotypes, the fraction of these that are local peaks is just the number of genotypes divided by the chance that any genotype is a local peak, hence the formula.) This little formula implies that random landscapes have hyperastronomical numbers of local peaks. For $N = 100$, there are about 10^{28} local peaks!

It begins to be obvious why adaptive search on random landscapes is very hard indeed. Suppose we wanted to find the highest peak. We try to search by climbing uphill. The adaptive walk soon becomes trapped on a local peak. The chance that the local peak is the highest peak, the global peak, is inversely proportional to the number of local peaks. So even for our modest genotype, with only 100 genes, not the 100,000 that humans have, the search process has about one chance in 10^{28} to climb to the highest peak. On random landscapes, finding the global peak by searching uphill is totally useless; we have to search the entire space of possibilities. But even for modestly complex genotypes, or programs, that would take longer than the history of the universe.

From any initial point on a landscape, adaptive walks reach local peaks after some number of steps (Figure 8.4b). Thus we can ask what are the expected lengths of such walks to peaks. On random landscapes, because there are so many peaks, the expected lengths of walks to peaks are very short (only ln N, where ln N is the logarithm of N taken with respect to the base e). (The logarithm of a number is simply the power some other number must be raised to in order to get the first number. Thus the base-10 logarithm of 1,000 is 3; of 100, 2.) Thus as N increases, say from 10 to 10,000, the number of genotypes increases enormously—indeed, exponentially—but expected walk lengths to optima increase only slightly (from about 2.3, the logarithm of 10, to about

9.2, the logarithm of 10,000). As in our moonscape image, any point is very close to local peaks that trap the adapting population and prevent further search for distant high peaks.

But the situation is even worse than that, for climbing toward the peaks rapidly becomes very much harder the higher one climbs. Figure 8.4*b* shows this central feature of adaptive walks. If one starts at a point of low fitness, some number of directions lead uphill. At each step uphill, the expected number of directions that continue uphill dwindles, until, at a local optimum, no directions are uphill. Is there some scaling law about how the fraction of directions uphill dwindle as the walk proceeds? On random landscapes, the answer is remarkable and simple. After each improvement step, the expected number of directions uphill is *cut in half*. If $N = 10,000$ and we start with the least fit, the expected number of directions uphill is, successively, 10,000 5,000, 2,500, 1,250 . . . So the higher one is, the harder it is to find a path that continues upward. At each step upward, one will have to try *twice* as many routes. Naturally, the expected waiting time to take a step uphill also doubles after each step uphill: the first step takes one try, the second takes two tries, then four, eight, 16. By the tenth step uphill, 1,024 tries are required. By the thirtieth step uphill, one must try 2^{30} directions to find a route uphill! This kind of slowing as fitness increases is a fundamental feature of all adaptive processes on even modestly rugged landscapes. This slowing, I will suggest in Chapter 9, underlies major features of biological and technological evolution.

Consider a further aspect of the plight of a population of organisms trying to evolve in such an unfriendly terrain. Random landscapes have very large numbers of local optima, one of which is the global optimum. If the population starts at a point and can climb in all possible alternative routes uphill, how many of the local peaks can be climbed from that initial point? On random landscapes, no matter where an adaptive walk starts, if the population is allowed to walk only uphill, it can reach only an infinitesimal fraction of the local peaks. Random landscapes are a bit like hitchhiking in New England: you can't get there from here. Try to evolve on a random landscape, seeking the highest possible peak, and the population remains boxed into infinitesimally small regions of the space of possibilities.

In these random moonscapes of impossible cliffs soaring and plummeting in every direction, no clues exist about where to go, only the confusion of a hyperastronomical number of local peaks, pinnacles studded into the vast space of possibilities. No pathways uphill lead farther than a few steps before the stunned and dazed hiker ascends to some minor peak, one of so many that the stars in the sky are a minor number by comparison. Start anywhere climbing uphill, and one will remain frozen in that tiny region of the space forever.

Correlated Fitness Landscapes

No complex entity that has evolved has done so on a random fitness landscape. The organisms out your window; the cells of which each is composed; the DNA, RNA, and protein molecules within those cells; the ecosystem of a forest, an alpine meadow, or a prairie; even the ecosystem of technologies in which we earn our livings—the standard operating procedures on the USS *Forrestal,* the linked production procedures in a General Motors plant, British common law, a telecommunications network—all evolve on landscapes in which minor "mutations" can cause both small and large variations.

Things capable of evolving—metabolic webs of molecules, single cells, multicellular organisms, ecosystems, economic systems, people—all live and evolve on landscapes that themselves have a special property: they allow evolution to "work." These real fitness landscapes, the types that underlie Darwin's gradualism, are "correlated." Nearby points tend to have similar heights. The high points are easier to find, for the terrain offers clues about the best directions in which to proceed. Unlike the jagged moonscape with cliffs plummeting and soaring from each perch, these landscapes may be smooth and flat like Nebraska, smooth and rounded like the gentle hills of Normandy, or even rugged like the Alps. Evolution can succeed on such landscapes. The Alps are staggering, but easy to scale compared with a random moonscape. With a compass, a backpack, a simple lunch, and good climbing gear, you can find your way to Mont Blanc and back in a day. A hard day, a wonderful day. A high-country adventure.

But if real landscapes are correlated, how do we go about studying them? We now confront a problem in building useful theories. A random landscape is almost uniquely defined: we merely assign random fitness values out of some distribution to genotypes in a genotype space. But what about correlated landscapes? For all we know, there may be indefinitely many ways to make correlated landscapes. Can we find a fruitful way?

I had no idea how to proceed until some years ago when new friends at the Santa Fe Institute—solid-state physicists Dan Stein from the University of Arizona, Richard Palmer from Duke, and Phil Anderson from Princeton—began talking about spin-glasses. Spin-glasses are a kind of disordered magnetic material, and Anderson was one of the first physicists to introduce models to understand their behavior. The *NK* model I will introduce is a kind of genetic version of a physicist's spin-glass model. The virtues of the *NK* model are that it specifically shows how different features of the genotype will result in landscapes with different degrees of ruggedness, and it allows us to study a family of landscapes in a controlled way.

Once again, we will be looking at organisms through the abstract lens of the theoretical biologist. I ask you to consider an organism with N traits, each having two alternative states: 0 and 1. These symbols might stand for "short nose" and "long nose" for one trait, and "bowlegged" versus "straight legged" for the other trait. Then a given organism is a unique combination of the 1 or 0 state of each of the N traits. By now it is obvious that there are 2^N possible combinations of traits, each a possible overall phenotype of our hypothetical organism. For example, we could have an organism with a short nose and bowlegs, a long nose and bowlegs, a short nose and straight legs, or a long nose and straight legs. We can designate these 00, 01, 10, and 11.

The fitness of an organism depends on which traits it has. So suppose we wish to understand the fitness of any one of these possible organisms, in terms of the specific combinations of its traits. Here is the problem: in a fixed environment, the contribution of one trait—say, short versus long nose—to the organism's fitness might depend on other traits—for example, bowed versus straight legs. Perhaps having a short nose is very useful if one is also bowlegged, but a short nose is harmful if one is straight legged. (More plausibly, thick bones may be useful for a massive organism, but harmful for one that is slender and fleet.)

In short, the contribution to overall fitness of the organism of one state of one trait may depend in very complex ways on the states of many other traits. Similar issues arise if we think of a haploid genotype with N genes, each having two alleles. The fitness contribution of one allele of one gene to the whole organism may depend in complex ways on the alleles of other genes. Geneticists call this coupling between genes epistasis or epistatic coupling, meaning that genes at other places on the chromosomes affect the fitness contribution of a gene at a given place. Thus there may be two genes, L and N, each with two alleles, L and l, as well as N and n. The L gene controls legs: L yields straight legs; l yields bowlegs. The N gene controls nose size: N yields big nose; n yields little nose. Since the alleles of the two genes control these traits, and the usefulness of a big nose may depend on leg shape, the fitness contribution of each allele of each gene to the organism's overall fitness may depend on the allele present at the other gene.

We can think of this phenomenon of genes affecting the fitness contributions of other genes as a network of epistatic interactions. Remember the genomic networks in Chapter 4, in which genes could switch one another on and off? Here we are after a slightly different idea. If we draw each gene as a node, we can connect each gene to all the genes that affect its fitness.

The *NK* model captures such networks of epistatic couplings and models the complexity of the coupling effects. It models epistasis itself by assigning to each trait, or gene, epistatic "inputs" from *K* other traits or genes. Thus the fitness contribution of each gene depends on the gene's own allele state, plus the allele states of the *K* other genes that affect that gene.

The actual epistatic effects between genes is very complex. One allele of a given gene may sharply enhance the fitness contribution of a given allele of a second gene, while a different allele of the first gene may cause the same allele of the second gene to be detrimental. Geneticists know that such epistatic effects occur, but difficult experiments are required to establish the details of such couplings between even two genes in an organism. Trying to establish all the epistatic effects among thousands of genes is, at present, unfeasible even in one species, let alone many species.

Several factors suggest that we might fruitfully model complex epistatic interactions by assigning the effects *at random* in some way. First, we may as well admit our ignorance in the biological cases at the present moment. Second, we are trying to build rather general models of rugged but correlated landscapes to begin to understand what such landscapes look like and what organismal features bear on landscape ruggedness. If we model the fitness effects of epistatic coupling "at random," we will obtain the kind of general models of landscapes we seek. Third, if we are lucky, we will find that real landscapes in some cases look very much like our model landscapes. We will have captured the right statistical features of real landscapes with our model correlated landscapes. So we may be able to understand the kind of epistasis that occurs in organisms and its effect on landscapes and evolution without needing to carry out all the experiments to establish all the details of epistatic couplings in any one organism. In short, we can seek general laws in biology concerning the structure and even the evolution of fitness landscapes by building models.

To complete an example of an *NK* fitness landscape is easy. Assign to each of the *N* genes *K* other genes. These might be chosen at random or in some other way. The fitness contribution of each gene to the whole genotype depends on *its own allele,* 1 or 0, *plus* the 1 or 0 alleles of the *K* other genes that are its inputs. Hence its fitness contribution depends on the alleles present in *K* + 1 genes. Since each gene can be in one of two allele states, 1 or 0, the total number of possible allele combinations is $2^{(K + 1)}$. We randomly assign each of these combinations of alleles a decimal number between 0.0 and 1.0 indicating how much fitness the gene confers on the organism. For one combination, the fitness contribution might be 0.76; for another, it might be 0.21. Now we must do the same thing for each of the N genes, and the $K + 1$ genes affecting it, randomly assigning each a different fitness contribution (Figure 8.5). It remains to consider the fitness of the entire genotype, which I shall define as the average fitness contribution of each of the N genes. To find how fit the entire organism is, add up the fitness contributions of each of the N genes and divide by N.

That's all there is to the NK model. Figure 8.5 shows an example with $N = 3$ and $K = 2$—that is, the genome has three genes, each of which is affected by two others. So, in this case, the fitness contribution of each gene depends on all the genes in the genome, itself plus the rest. K, in other words, has its maximal value, $N - 1$.

The NK model in Figure 8.5 yields a fitness landscape for each of the 2^3, or eight, possible genotypes, each located at the corner of a three-dimensional Boolean cube: (000), (001) . . . (111). Note that all our ideas about such landscapes are visible: there are two local optima that can be

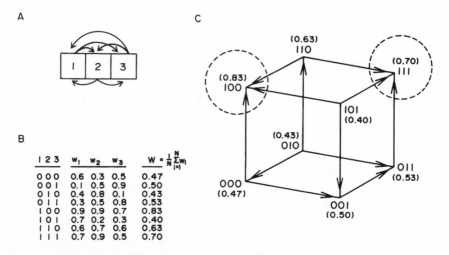

Figure 8.5 *Building a fitness landscape. The NK model of a genomic network consisting of three genes (N = 3), each of which can be in one of two states, 1 or 0. Each gene receives input from two other genes (K = 2). (a) Two inputs are arbitrarily assigned to each gene. (b) Each gene in each of the 2^3 = 8 possible genomes is randomly assigned a fitness contribution between 0.0 and 1.0. Then the fitness of each genome is computed as the mean value of the fitness contributions of the three genes. (c) A fitness landscape is constructed. Circled vertices again represent local optima.*

reached with various walks; there are dwindling numbers of directions uphill along such walks; and so forth.

I find the *NK* model fascinating because of this essential point: altering the number of epistatic inputs per gene, *K*, alters the ruggedness and number of peaks on the landscape. Altering *K* is like twisting a control knob. Why does this happen? Because our model organism, with its network of epistatic interactions among its genes, is caught in a web of conflicting constraints. The higher *K* is—the more interconnected the genes are—the more conflicting constraints exist, so the landscape becomes ever more rugged with ever more local peaks.

It is easy to see why increasing *K* increases conflicting constraints. Suppose that two genes share many of the same *K* inputs. The fitness contribution of each combination of allele states has been assigned at random. Thus, almost certainly, the best choice of the allele states of the shared epistatic inputs for gene 1 will be different than for gene 2. There are conflicting constraints. There is no way to make gene 1 and gene 2 as "happy" as they might be if there were no cross-coupling between their epistatic inputs. Thus as *K* increases, the cross-coupling increases and the conflicting constraints become ever worse.

It is these conflicting constraints that make the landscape rugged and multipeaked. Because so many constraints are in conflict, there is a large number of rather modest compromise solutions rather than an obvious superb solution. There are, in other words, many local peaks with very low altitudes. Because landscapes are more rugged, adaptation becomes harder. Recalling our compressed computer program, increasing *K* is like increasing the compression of the program. In both cases, as this occurs, each small part of the system affects other parts of the whole system. As the density of interconnections increases, changing a single gene (or bit in the program) will have effects that ripple throughout the system. It will be difficult indeed for it to smoothly evolve, with slight changes in the genome making correspondingly slight changes in fitness. So putting the effects together, as *K* increases, the heights of peaks decrease, their numbers increase, and evolving over the landscape becomes more difficult.

To sharpen our intuitions about the structure of landscapes and their effects on evolvability, let us start with our knob set at *K* = 0, so that each gene is independent of every other gene. There are no conflicting constraints, because there are no epistatic inputs, no cross-connections. The landscape is a "Fujiyama" landscape, with a single peak falling away along smooth gradual slopes. This is easy to understand. Suppose, without loss of generality, that the "1" value happened to be the fitter allele for each gene. Then there is a unique genotype (1111111111), which is obviously the global optimum. But any other genotype, say

(0001111111), could obviously climb to the global optimum by "flipping" successively each of the 0 alleles to the 1 alleles. So there are no other peaks on the landscape, for any other genotype can climb to the global optimum. But further, the fitness of near neighbors cannot be very different, since changing a single gene from 1 to 0 can change genotype fitness by at most only $1/N$. So the sides of the peak are smooth. The single peak is far away from a typical starting point. If one started with a random genotype, half the genes would be expected to be 0, and the remaining would be 1; thus the expected walk length or distance to the peak is $N/2$ mutational steps, halfway across the space. And at each step uphill, the number of directions uphill dwindles by only 1. If one started with the worst genotype, there would be N directions uphill, then $N - 1$, then $N - 2$, and so forth along an adaptive walk leading inevitably to the global optimum. This gradual dwindling of the number of directions uphill is in sharp contrast to random landscapes where the number falls by half at each step uphill. On such a smooth, single-peaked landscape, a population of adapting organisms randomly mutating genes, and selecting the fitter genotype, would quickly find its way to the top of Fujiyama. Here is the ideal gradualism of Darwin.

But suppose that the K knob is turned up the other way to its maximum value, $N - 1$, so that every gene is affected by every other gene. When K increases to its maximal value, $N - 1$, the fitness landscape is completely random (Figure 8.5). This is easy to see. Changing any single gene to the other allele affects that gene and all other genes. For each affected gene, its fitness contribution is switched to a new random value between 0.0 and 1.0. Since this is true for all N genes, the fitness of the new genotype, mutated in only a single gene, is completely random with respect to the initial genotype.

Since $K = N - 1$ landscapes are fully random, all the properties we have noted arise. The landscape has $2^N/(N + 1)$ optima, so that when the number of genes is large, the number of local peaks is hyperastronomical. Walks to optima are very short. Starting anywhere, an adapting system can climb uphill to only an infinitesimal fraction of the local optima, meaning that an adapting system is frozen into a tiny region of its state space. With every step uphill, the number of directions uphill dwindle by half, so the rate of improvement slows rapidly as the system reaches higher and higher fitness. For all these reasons, adaptive evolution to the highest peaks is virtually impossible.

As K is tuned from its minimum value, 0, to its maximum value, $N - 1$, a family of correlated but increasingly rugged landscapes is generated. Landscapes become increasingly rugged and multipeaked, while the peaks become lower. This can be seen in Figure 8.6, which shows the landscape in the immediate neighborhood of a local peak for increasing

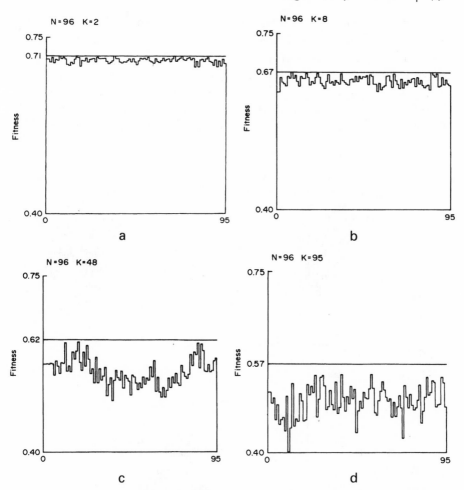

Figure 8.6 *Tuning ruggedness. The ruggedness of a fitness landscape can be tuned by adjusting the number of inputs, K, per gene. Here the landscape in the immediate vicinity of a local optimum is plotted for increasing values of K. From (a) to (d), K gets higher; the peaks decrease in fitness, while the neighborhood around the peak becomes more rugged.*

values of K. The peaks get lower as K increases, while the neighborhood of the peak becomes more rugged. My aim in inventing the NK model was to allow study of such rugged but correlated landscapes.

Evolving on Rugged Landscapes

Real landscapes are neither as simple as the Fujiyama landscape nor fully random. All organisms—and all kinds of complex systems—evolve on correlated landscapes tuned somewhere between the Fujiyama $K = 0$ landscape and $K = N - 1$ moonscape—landscapes that are rugged but not random. So now we must ask if there are general features of these rugged landscapes that can deepen our insight into how evolution works.

Using a computer simulation, we can take a God's-eye view of NK landscapes, gazing down on the surprising large-scale features below. Figure 8.7 shows that when K is low, peaks cluster near one another like

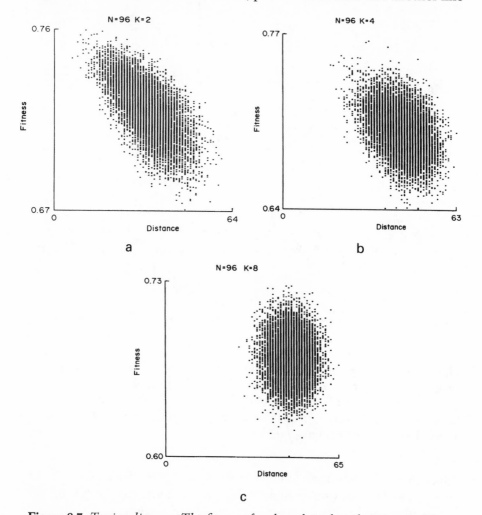

a

b

c

Figure 8.7 *Tuning distance. The fitness of each peak is plotted against its "distance" from the highest peak found. (a) If the number of epistatic inputs is low, $K = 2$, high peaks cluster near one another, creating an oblong shape oriented from the upper left to the lower right. (b) As K increases to 4, high peaks begin to spread apart from one another and the oblong is more nearly vertical. (c) When K reaches 8, high peaks have no tendency to be near one another and the oblong is vertical.*

the high peaks in the Alps. With few conflicting constraints, the landscape is *nonisotropic:* there is a special region where the high peaks cluster. Therefore, it is useful to an adaptive process to locate this special region in the space of possibilities. But as *K* increases and the landscapes become ever more rugged, the high peaks spread apart from one another over the landscape. So on very rugged landscapes, the high peaks are dotted randomly across the landscape. As this occurs, the landscape becomes *isotropic:* any area is about the same as any other area. In turn, this implies that there is no point in searching far away on an isotropic landscape for regions of good peaks; such regions simply do not exist.

Turning our attention to Figure 8.8, we see that landscapes with

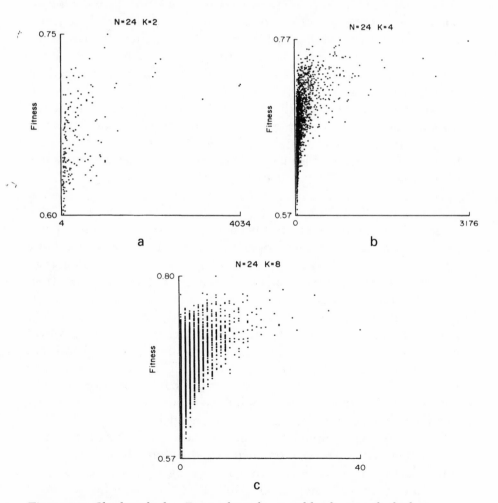

Figure 8.8 *Climbing higher. For moderately rugged landscapes, the highest peaks are the ones that can be scaled from the greatest number of initial positions. This is shown for three different values:* (a) K = 2, (b) K = 4, (c) K = 8.

moderate degrees of ruggedness share a striking feature: it is the highest peaks that can be scaled from the greatest number of initial positions! This is very encouraging, for it may help explain why evolutionary search does so well on this kind of landscape. On a rugged (but not random) landscape, an adaptive walk is more likely to climb to a high peak than a low one. If an adapting population were to "jump" randomly into such a landscape many times and climb uphill each time to a peak, we would find that there is a relationship between how high the peak is and how often the population climbed to it. If we turned our landscapes upside down and sought instead the lowest valleys, we would find that the *deepest valleys drain the widest basins.*

The property that the highest peaks are the ones to which the largest fraction of genotypes can climb is not inevitable. The highest peaks could be very narrow but very high pinnacles on a low-lying landscape with modest broad hilltops. If an adapting population were released at a random spot and walked uphill, it would then find itself trapped on the top of a mere local hilltop. The exciting fact we have just discovered is that for an enormous family of rugged landscapes, the *NK* family, the highest peaks "drain" the largest basins. This may well be a very general property of most rugged landscapes reflecting complex webs of conflicting constraints. So this may be a very general property of the kinds of landscapes undergirding biological (and technological) evolution.

Recall another striking feature of random landscapes: with every step one takes uphill, the number of directions leading higher is cut by a constant fraction, one-half, so it becomes ever harder to keep improving. As it turns out, the same property shows up on almost any modestly rugged or very rugged landscape. Figure 8.9 shows the dwindling fraction of fitter neighbors along adaptive walks for different *K* values (Figure 8.9a) and the increased waiting times to find fitter variants for different *K* values (Figure 8.9b). Once *K* is modestly large, about $K = 8$ or greater, at each step uphill the number of directions uphill falls by a *constant fraction,* and the waiting time or number of tries to find that way uphill increases by a constant fraction. This means that as one climbs higher and higher, it becomes not just harder, but *exponentially harder* to find further directions uphill. So if one can make one try per unit time, the rate of improving slows exponentially.

If one is struggling uphill on a random landscape, looking for ways to improve, at first it takes one try, then two, then four, then eight, and so forth, the number of directions uphill dropping by half at each step so the number of tries to step uphill doubles. So the rate of improvement slows exponentially. On somewhat smoother landscapes, with *K* modestly large, the number of directions uphill drops by less than half at

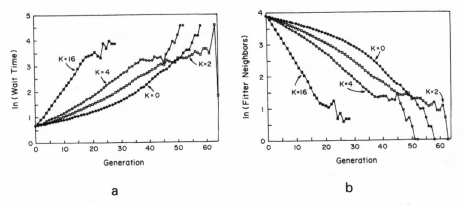

Figure 8.9 *Diminishing returns. On rugged landscapes, for every step taken uphill (toward higher fitness) the number of paths leading still higher is cut by a constant fraction and the number of fitter neighbors decreases. (a) The natural logarithm of the waiting time, or number of tries, to find a fitter neighbor at each adaptive step (generation) is shown for different values of K. (b) The natural logarithm of the fraction of fitter neighbors at each step is shown for different values of K.*

each step. But the rate of slowing is again exponential. As epistatic cross-coupling, K, increases, the landscape becomes more rugged and adaptation slows ever more rapidly.

We have just seen an utterly fundamental property of very many rugged fitness landscapes. The rate of improvement slows exponentially as one clambers uphill. This is no mere abstraction. We will find that such exponential slowing is characteristic of technological as well as biological evolution.

Why? Because both biological evolution and technological evolution are processes attempting to optimize systems riddled with conflicting constraints. Organisms, artifacts, and organizations evolve on correlated but rugged landscapes. Suppose we are designing a supersonic transport and have to place the fuel tanks somewhere, but also have to design strong but flexible wings to carry the load, install controls to the flight surfaces, place seating, hydraulics, and so forth. Optimal solutions to one part of the overall design problem conflict with optimal solutions to other parts of the overall design. Then we must find *compromise* solutions to the joint problem that meet the conflicting constraints of the different subproblems. Likewise, a foraging organism must allocate time and resources to foraging activities. But speed over its turf is in conflict with careful examination of each spot for food. How should these conflicting requirements be jointly optimized? A tree may utilize metabolic resources to make chemical toxins to ward off insects, rather than utilize the same resources to build leaves to capture sunlight. How should the tree solve the conflicting constraints in its budget allocation?

Organisms and artifacts evolve on rugged landscapes as a result of such conflicting constraints. In turn, those conflicting constraints imply that as adaptive searches climb higher toward the peaks, the rate of improvement slows exponentially.

A God's-Eye View

It is easy enough for us, the creators of these models, to gaze down on the peaks and valleys and divine their large-scale features. But what of the organisms themselves? In all the examples we have discussed so far, the evolving population is limited by the fact that the only clues it can garner about how to move higher on the fitness landscape are the ones right in front of its nose. Evolution by mutation and selection alone is limited to searching locally in the space of possibilities, guided only by the local terrain.

If only an adapting population could take a God's-eye view and behold the large-scale landscape features—see where to evolve rather than just climbing blindly uphill from its current positions, only to become trapped on poor local peaks. If only an adapting population could see beyond its nose.

Ah, but it can. And sex is probably the answer—a kind of genetic variation we have not yet considered.

Why sex has evolved is one of the great mysteries of biology. You see, if you are a happy bacterium, dividing away, your fitness is just your own rate of division. But as soon as sexuality evolves, it takes two— mother and father—to create a single offspring. There is an immediate twofold loss in fitness! If it takes two parents to produce a single offspring, why bother? Because sex has evolved, biologists think, to permit genetic recombination. And recombination provides a kind of approximation to a God's-eye view of just these large-scale features of fitness landscapes.

Sexual organisms are diploids, not haploids. Recall that the egg will contain half of the mother's diploid chromosomes, and the sperm will contain half of the father's diploid chromosomes. These will be passed to the zygote, re-creating the total diploid number of chromosomes. During maturation of the egg cell or of the sperm cell, the process of meiosis occurs. A random half of the mother's own parental chromosomes will be passed to the egg cell. A random half of the father's own parental chromosomes will be passed to the sperm cell. But before a random half of the parental chromosomes are chosen, recombination can occur between the maternal and paternal chromosomes in the cell giving rise to the egg or in the cell giving rise to the sperm. Take the case of the cell giving rise to an egg. That cell has maternal and paternal copies of a number of different chromosomes. Each maternal and paternal pair lines up next to each other. With recombination, a break occurs at the same position in both chromosomes, and the left end of the paternal chromosome is joined to the right end of the maternal chromosome to create one new recombined chromosome, while the right end of the paternal chromosome is joined to the left end of the maternal chromosome to create the second recombined chromosome. If the alleles on the two chromosomes are (000000) and (111111) and a break occurs between the second and third genes, then the two recombinants might be (001111) and (110000).

How does sex approximate a God's-eye view of a landscape? Suppose that an adapting population is spread out over some region of a fitness landscape. Recombination between organisms at different locations on the landscape allows the adapting population to "look at" the regions *between* the parental genotypes. Suppose that the maternal and paternal chromosomes were (111100) and (111111); then recombination would create no new patterns in the first four genes, but could create the two novel combinations (111101) and (111110) of the last two genes on single chromosomes. Here recombination is sampling between the two parental genotypes. Suppose that the two parental chromosomes were entirely different, (111111) and (000000). Recombination could occur at any point, between the first and second genes, between the second and third, and so forth. So recombination could create a large number of different genotypes, (111110), (111100) . . . (000001), all of which lie between the two parental genotypes in genotype space.

Recombination allows an adapting population to make use of large-scale features of the landscape to find high peaks. Those same large-scale features would be relatively or fully invisible to a haploid population climbing uphill, guided only by local clues.

If the landscape looks like the Alps, the highest peaks cluster near one another. Thus the locations of high peaks carry mutual information about the locations of still higher peaks. If you are on or near a high peak and I am on or near another, the region *between* us is a fine place to look for still higher peaks! If you are in Geneva and I am in Milan, the region between us is certainly a good place to look for good skiing.

If the highest peaks "drain" the largest basins, then recombination can also make use of this large-scale feature. If you and I are well up on the sides of high peaks, get married, and, thanks to recombination, drop off our offspring at random locations between us, the kids will have a very good chance to land at spots already of high fitness—and, more important, a spot from which they can climb to still higher peaks!

To unpack the metaphor, if a population uses simple mutation and selection to climb toward peaks in its vicinity, and also uses recombination to explore the genotype space between the members of the population, it can utilize both local and large-scale features of the fitness landscape to adapt. Fitness can increase far more rapidly. Just this occurs in our *NK* landscapes; populations adapting using both mutation and recombination as well as selection improve far more readily than those using only mutation and selection.

No wonder most species are sexual. But our puzzle is only half stated. Recombination is a dazzling search tactic on some fitness landscapes, but a disaster on others. For example, recombination is useless on a random landscape. It is worse than useless. If you and I have climbed to local peaks, at least we are at local peaks. If we fiddle with recombination, our progeny are dropped off in no-man's-land and will, on average, have much, much lower fitness than we do. Recombination is positively harmful on the "wrong" kind of fitness landscapes.

So if most species are sexual, with the twofold cost in fitness entailed by the bother, it must be the case that recombination is generally useful. Therefore, either the kinds of conflicting constraints that occur in organisms just happen to yield fitness landscapes where recombination is useful or selection itself has "opted for" or constructed organisms having the property that recombination is useful.

Which of these alternatives is correct? I do not know. I'd bet a mixture of both. We again begin to get a glimpse of the mystery of evolution. Selection, in crafting the kinds of organisms that exist, may also help craft the kinds of landscapes over which they evolve, picking landscapes that are most capable of supporting evolution—not only by mutation alone, but by recombination as well. Evolvability itself is a triumph. To benefit from mutation, recombination, and natural selection, a population must evolve on rugged but "well-correlated" landscapes. In the framework of *NK* landscapes, the "*K* knob" must be well tuned. Either we are very, very lucky and happen to have the density of epistatic connections that lets us evolve, or there was something that adjusted the knob. Those who believe that natural selection is the sole source of biological order must assume that selection itself achieved organisms with the proper level of epistasis, the right value of K. But is selection powerful enough to craft the structure of fitness landscapes? Or are there limits to the power of selection? If selection is too weak to ensure evolvability, how is such evolvability achieved and maintained? Could self-organization play a role? Here is a profound problem that must re-shape our thinking.

The Limits to Selection

If selection could, in principle, accomplish "anything," then all the order in organisms might reflect selection alone. But, in fact, there are limits to selection. Such limits begin to demand a shift in our thinking in the biological sciences and beyond.

We have already encountered a first powerful limitation on selection. Darwin's view of the gradual accumulation of useful variations, we saw, required gradualism. Mutations must cause slight alterations in pheno-types. But we have now seen two alternative model "worlds" in which such gradualism fails. The first concerns maximally compressed pro-grams. Because these are random, almost certainly any change random-izes the performance of the program. Finding one of the few useful minimal programs requires searching the entire space—requiring un-thinkably long times compared with the history of the universe even for modestly large programs. Selection cannot achieve such maximally compressed programs. Our second examples are NK landscapes. If the richness of epistatic couplings, K, is very high and approaches the maxi-mum limit, $K = N - 1$, landscapes approach and become completely random. Again, locating the highest peak or one of the few highest peaks requires searching the entire space of possibilities. For modestly large genomes, this is simply impossible.

But the matter is even worse on such random landscapes. If an adapt-ing population evolves by mutation and selection alone, it will remain frozen in an infinitesimal region of the total space, trapped forever in whatever region it started in. It will be unable to search long distances across space for the high peaks. Yet if the population dares try recombi-nation, it will be harmed, on average, not helped.

There is a second limitation to selection. It is not only on random landscapes that selection fails. Even on smooth landscapes, in the heart-land of gradualism, just where Darwin's assumptions hold, selection can again fail and fail utterly. Selection runs headlong into an "error ca-tastrophe" where all accumulated useful traits melt away.

Let us return to our image of a population of bacteria evolving on a rugged fitness landscape. The behavior of the population depends on the size of the population, the mutation rate, and the structure of the landscape. Suppose we consider holding population size constant—say, by using a chemostat—and landscape structure constant, and tuning the mutation rate from low to high by some experimental technique. What will happen? Suppose the population is initially genetically identical; hence all bacteria are located at the same point in genotype space. If the mutation rate is very low, then at long intervals a fitter variant arises and rapidly sweeps through the population. Thus the population as a whole "hops" to the fitter neighboring genotype. Over time, the population performs just the kind of adaptive walk we have considered, climbing steadily uphill to some local optimum and remaining there.

But what happens if the mutation rate is so high that many fitter and less fit variants are found over very short intervals? Then the population will spread out from the initial point in genotype space and climb in many directions. The more surprising property is this: even if the population is released on a local peak, it may not stay there! Simply put, the rate of mutation is so high that it causes the population to "diffuse" away from the peak faster than the selective differences between less fit and more fit mutants can return the population to the peak. An error catastrophe, first discovered by Nobel laureate Manfred Eigen and theoretical chemist Peter Schuster, has occurred, for the useful genetic information built up in the population is lost as the population diffuses away from the peak.

To summarize: as the mutation rate increases, at first the population climbs a local hill and hovers in its vicinity. As the mutation rate becomes higher, the population drifts down from the peak and begins to spread along ridges of nearly equal fitness on the fitness landscape. But if the mutation rate increases still further, the population drifts ever lower off the ridges into the lowlands of poor fitness.

Eigen and Schuster were the first to emphasize the importance of this error catastrophe, for it implies a limit to the power of natural selection. At a high enough mutation rate, an adapting population cannot assemble useful genetic variants into a working whole; instead, the mutation-induced "diffusion" over the space overcomes selection, pulling the population toward adaptive peaks.

This limitation is even more marked when seen from another vantage point. Eigen and Schuster also emphasized that for a constant mutation rate per gene, the error catastrophe will arise when the number of genes in the genotype increases beyond a critical number. Thus there appears to be a *limit on the complexity* of a genome that can be assembled by mutation and selection!

Selection, then, confronts twin limitations: it is trapped or frozen into local regions of very rugged landscape, and, on smooth landscapes, it suffers the error catastrophe and melts off peaks, so the genotype becomes less fit. Selection may not be powerless in the face of these limitations, for selection can play a role in molding the ruggedness of the landscapes over which organisms evolve. We have already seen some of the ways, for the NK model itself shows that molding the epistatic interactions among genes modifies ruggedness. But since the power of selection is limited, it seems doubtful that selection alone can ensure friendly landscapes. Perhaps another source of order is required. Evolution may be impossible without the privilege of working with systems that already exhibit internal order, with fitness landscapes already naturally tuned so that natural selection can get a foothold and do its job.

And here, I think, may be an essential tie between self-organization and selection. Self-organization may be the *precondition* of evolvability itself. Only those systems that are able to organize themselves spontaneously may be able to evolve further. How far we have come from a simple picture of selection merely sifting for fitter variants. Evolution is far more subtle and wonderful.

Self-Organization, Selection, and Evolvability

Whence the order out my window? Self-organization *and* selection, I think. We, the expected, *and* we, the ad hoc. We, the children of ultimate law. We, the children of the filigrees of historical accident.

What is the weave? No one yet knows. But the tapestry of life is richer than we have imagined. It is a tapestry with threads of accidental gold, mined quixotically by the random whimsy of quantum events acting on bits of nucleotides and crafted by selection sifting. But the tapestry has an overall design, an architecture, a woven cadence and rhythm that reflect underlying law—principles of self-organization.

How are we to begin to understand this new union? For "begin to understand" is all we can now hope for. We enter new territory. It would be presumptuous to suppose that we would understand a new continent when first alighting on its nearest shores. We are seeking a new conceptual framework that does not yet exist. Nowhere in science have we an adequate way to state and study the interleaving of self-organization, selection, chance, and design. We have no adequate framework for the place of law in a historical science and the place of history in a lawful science.

But we are beginning to pick out themes, strands in the tapestry. The first theme is self-organization. Whether we confront lipids spontaneously forming a bilipid membrane vesicle, a virus self-assembling to a low-energy state, the Fibonacci series of a pinecone's phyllotaxis, the emergent order of parallel-processing networks of genes in the ordered regime, the origin of life as a phase transition in chemical reaction systems, the supracritical behavior of the biosphere, or the patterns of co-evolution at higher levels—ecosystems, economic systems, even cultural systems—we have found the signature of law. All these phenomena give signs of nonmysterious but emergent order. We begin to believe in this new strand, to sense its power. The problems are twofold: first, we do not yet understand the wealth of sources of such spontaneous order; second, we have the gravest difficulties understanding how self-organization might interact with selection.

Selection is the second theme. Selection is no more mysterious than self-organization. I hope I have persuaded you that selection is powerful, but limited. It is not the case that all complex systems can be assembled by an evolutionary process. We must try to understand what kinds of complex systems can actually arise this way.

The inevitability of historical accident is the third theme. We can have a rational morphology of crystals, because the number of space groups that atoms in a crystal can occupy is rather limited. We can have a periodic table of the elements because the number of stable arrangements of the subatomic constituents is relatively limited. But once at the level of chemistry, the space of possible molecules is vaster than the number of atoms in the universe. Once this is true, it is evident that the actual molecules in the biosphere are a tiny fraction of the space of the possible. Almost certainly, then, the molecules we see are to some extent the results of historical accidents in this history of life. History arises when the space of possibilities is too large by far for the actual to exhaust the possible.

Stating the themes is simple. It is their interleaving that is so terribly uncertain.

Here is a firm foothold: an evolutionary process, to be successful, requires that the landscapes it searches are more or less correlated. What kinds of real physico-chemical systems exhibit correlated landscapes, the gradualism Darwin assumed? While I have no exhaustive answer, we already have clues. Our lipid vesicle is stable in the sense Darwin requires. Many minor modifications of the molecular structure of the lipids, of the mixture of lipids, of lipid and nonlipid molecules, of the medium, all leave a lipid vesicle fundamentally intact. Such a vesicle is in a stable low-energy equilibrium state. (In Chapter 1, we used the image of a ball rolling to the bottom of a bowl.) The stable morphology can be sustained, at least approximately, with respect to detailed alterations. The same holds true for a self-assembling virus, for the double helix of DNA or RNA, or for the folded proteins encoded by those genes. Cells and organisms make full use of such stable low-energy structures. We will be correct if we think of the stable structures such systems form as "robust."

Nonequilibrium systems can be robust as well. A whirlpool dissipative system is robust in the sense that a wide variety of shapes of the container, flow rates, kinds of fluids, and initial conditions of the fluids lead to vortices that may persist for long periods. So small changes in the construction parameters of the system, and initial conditions, lead to small changes in behavior.

Whirlpools are attractors in a dynamical system. Attractors, however, can be both stable and unstable. Instability arises in two senses. First, small changes in the construction of the system may dramatically alter the behavior of the system. Such systems are called structurally unstable. In addition, small changes in initial conditions, the butterfly effect, can sharply change subsequent behavior. Conversely, stable dynamical systems can be stable in both senses. Small changes in construction may typically lead to small changes in behavior. The system is structurally stable. And small changes in initial conditions can lead to small changes in behavior. The butterfly is asleep.

We have examined dynamical systems that fall into both the unstable and the stable categories. Large Boolean networks, our model genomic regulatory systems, can lie in the chaotic regime or the ordered regime, or can lie near the phase transition, in the complex regime at the edge of chaos.

We know that there is a clear link between the stability of the dynamical system and the ruggedness of the landscape over which it adapts. Chaotic Boolean networks, and many other classes of chaotic dynamical systems, are structurally unstable. Small changes wreak havoc on their behavior. Such systems adapt on very rugged landscapes. In contrast, Boolean networks in the ordered regime are only slightly modified by mutations to their structure. These networks adapt on relatively smooth fitness landscapes.

We know from the *NK* landscape models discussed in this chapter that there is a relationship between the richness of conflicting constraints in a system and the ruggedness of the landscape over which it must evolve. We plausibly believe that selection can alter organisms and their components so as to modify the structure of the fitness landscapes over which those organisms evolve. By taking genomic networks from the chaotic to the ordered regime, selection tunes network behavior to be sure. By tuning epistatic coupling of genes, selection also tunes landscape structure from rugged to smooth. Changing the level of conflicting constraints in the construction of an organism from low to high tunes how rugged a landscape such organisms explore.

Not only do organisms evolve, but, we must suppose, the structure of the landscapes that organisms explore also evolves. Since selection faces an error catastrophe on very smooth landscapes and can become excessively trapped in small regions of the space of possibilities on very rugged landscapes, we must also begin to suspect that selection seeks "good" landscapes. We do not as yet know in any detail what kinds of landscapes are "good," although it seems safe to conclude that such landscapes must be highly correlated, not random.

However, the very limits on selection we have discussed must raise questions about whether selection itself can achieve and sustain the kinds of organisms that adapt on the kinds of landscapes where selection works well. It is by no means obvious that selection can, of its own accord, achieve and sustain evolvability. Were cells and organisms not inherently the kinds of entities such that selection could work, how could selection gain a foothold? After all, how could evolution itself bring evolvability into existence, pulling itself up by its own bootstraps?

And so we return to a tantalizing possibility: that self-organization is a prerequisite for evolvability, that it generates the kinds of structures that can benefit from natural selection. It generates structures that can evolve gradually, that are robust, for there is an inevitable relationship among spontaneous order, robustness, redundancy, gradualism, and correlated landscapes. Systems with redundancy have the property that many mutations cause no or only slight modifications in behavior. Redundancy yields gradualism. But another name for redundancy is robustness. Robust properties are ones that are insensitive to many detailed alterations. The robustness of the lipid vesicle, or of the cell type attractors in genomic networks in the ordered regime, is just another version of redundancy. Robustness is precisely what allows such systems to be molded by gradual accumulation of variations. Thus another name for redundancy is structural stability—a folded protein, an assembled virus, a Boolean network in the ordered regime. The stable structures and behaviors are ones that can be molded.

If this view is roughly correct, then precisely that which is self-organized and robust is what we are likely to see preeminently utilized by selection. Then there is no necessary and fundamental conflict between self-organization and selection. These two sources of order are natural partners. The cell membrane is a bilipid membrane, stable for almost 4 billion years both because it is robust and because such robust forms are readily malleable by natural selection. The genomic network, I believe, lies in the ordered regime, perhaps near the edge of chaos, because such networks are readily formed, part of order for free, but also because such systems are structurally and dynamically stable, so they adapt on correlated landscapes and are able to be molded for further tasks.

But if selection has built organisms utilizing the properties that are self-organized and robust—both because those features lie to hand in evolution, and because the same self-organized features are just those which are readily crafted—then we are not merely tinkered-together contraptions, ad hoc molecular machines. The building blocks of life at a variety of levels from molecules to cells to tissues to organisms are precisely the robust, self-organized, and emergent properties of the way the world works. If selection merely molds further the stable properties of its building blocks, the emergent lawful order exhibited by such systems will persist in organisms. The spontaneous order will shine through, whatever selection's further siftings.

Can selection have reached beyond the spontaneous order of its building blocks? Perhaps. But we do not know how far. The more rare and improbable the forms that selection seeks, the less typical and robust they are and the stronger will be the pressure of mutations to revert to what is typical and robust. That natural order, we may suspect, will indeed shine through.

And so we are the locus of law. Evolution is surely "chance caught on the wing," but it is also the expression of underlying order.

We, the expected. We, at home in the universe.

Chapter 9

Organisms and Artifacts

Organisms arise from the crafting of natural order and natural selection, artifacts from the crafting of *Homo sapiens.* Organisms and artifacts so different in scale, complexity, and grandeur, so different in the time scales over which they evolved, yet it is difficult not to see parallels.

Life spreads through time and space in branching radiations. The Cambrian explosion is the most famous example. Soon after multicelled forms were invented, a grand burst of evolutionary novelty thrust itself outward. One almost gets the sense of multicellular life gleefully trying out all its possible ramifications, in a kind of wild dance of heedless exploration. As though filling in the Linnean chart from the top down, from the general to the specific, species harboring the different major body plans rapidly spring into existence in a burst of experimentation, then diversify further. The major variations arise swiftly, founding phyla, followed by ever finer tinkerings to form the so-called lower taxa: the classes, orders, families, and genera. Later, after the initial burst, after the frenzied party, many of the initial forms became extinct, many of the new phyla failed, and life settled down to the dominant designs, the remaining 30 or so phyla, Vertebrates, Arthropods, and so forth, which captured and dominated the biosphere.

Is this pattern so different from technological evolution? Here human artificers make fundamental inventions. Here, too, one witnesses, time after time, an early explosion of diverse forms as the human tinkerers try out the plethora of new possibilities opened up by the basic innovation. Here, too, is an almost gleeful exploration of possibilities. And, after the party, we settle down to finer and finer tinkering among a few dominant designs that command the technological landscape for some time—until an entire local phylogeny of technologies goes extinct. No one makes Roman siege engines any more. The howitzer and short-range rocket have driven the siege engine extinct.

Might the same general laws govern major aspects of biological and technological evolution? Both organisms and artifacts confront conflicting design constraints. As shown, it is those constraints that create rugged fitness landscapes. Evolution explores its landscapes without the benefit of intention. We explore the landscapes of technological opportunity with intention, under the selective pressure of market forces. But if the underlying design problems result in similar rugged landscapes of conflicting constraints, it would not be astonishing if the same laws governed both biological and technological evolution. Tissue and terracotta may evolve by deeply similar laws.

In this chapter, I will begin to explore the parallels between organism and artifact, but the themes will persist throughout the remainder of the book. I will explore two features of rugged but correlated landscapes. The first feature accounts, I believe, for the general fact that fundamental innovations are followed by rapid, dramatic improvements in a variety of very different directions, followed by successive improvements that are less and less dramatic. Let's call this the "Cambrian" pattern of diversification. The second phenomenon I want to explore is that after each improvement the number of directions for further improvement falls by a constant fraction. As we saw in Chapter 8, this yields an exponential slowing of the rate of improvement. This feature, I believe, accounts for the characteristic slowing of improvement found in many technological "learning curves" as well as in biology itself. Let's call this the "learning-curve" pattern. Both, I think, are simple consequences of the statistical features of rugged but correlated landscapes.

Jumping Across Landscapes

In our current efforts, I will continue to use the *NK* model of correlated fitness landscapes introduced in Chapter 8. It is one of the first mathematical models of tunably rugged fitness landscapes. I believe, but do not know, that the features we will explore here will turn out to be true of almost any family of rugged but correlated landscapes. As we have seen, the *NK* model generates a family of increasingly rugged landscapes as K, the number of "epistatic" inputs per "gene," increases. Recall that increasing K increases the conflicting constraints. In turn, the increase in conflicting constraints makes the landscape more rugged and multipeaked. When K reaches its maximal value ($K = N - 1$, in which every gene is dependent on every other), the landscape becomes fully random.

I begin by describing a simple, idealized kind of adaptive walk—long-jump adaptation—on a correlated but rugged landscape. We have already looked at adaptive walks that proceed by generating and selecting single mutations that lead to fitter variants. Here, an adaptive walk proceeds step-by-step in the space of possibilities, marching steadfastly uphill to a local peak. Suppose instead that we consider simultaneously making a large number of mutations that alter many features at once, so that the organism takes a "long jump" across its fitness landscape. Suppose we are in the Alps and take a single normal step. Typically, the altitude where we land is closely correlated with the altitude from which we started. There are, of course, catastrophic exceptions; cliffs do occur here and there. But suppose we jump 50 kilometers away. The altitude at which we land is essentially uncorrelated with the altitude from which we began, because we have jumped beyond what is called the *correlation length* of the landscape.

Now consider *NK* landscapes for modest values of *K*, say *N* = 1,000 and *K* = 5—1,000 genes whose fitness contributions each depends on 5 other genes. The landscape is rugged, but still highly correlated. Nearby points have quite similar fitness values. If we flip one, five, or 10 of the 1,000 genes, we will end up with a combination that is not radically different in fitness from the one with which we began. We have not exceeded the correlation length.

NK landscapes have a well-defined correlation length. Basically, that length shows how far apart points on the landscape can be so that knowing the fitness at one point still allows us to predict something about the fitness at the second point. On *NK* landscapes, this correlation falls off exponentially with distance. Therefore, if one jumped a long distance away, say changing 500 of the 1,000 allele states—leaping halfway across the space—one would have jumped so far beyond the correlation length of the landscape that the fitness value found at the other end would be totally random with respect to the fitness value from which one began.

A very simple law governs such long-jump adaptation. The result, exactly mimicking adaptive walks via fitter single-mutant variants on random landscapes is this: every time one finds a fitter long-jump variant, the expected number of tries to find a *still better* long-jump variant doubles! This simple result is shown in Figure 9.1. Figure 9.1*a* shows the results of long-jump adaptation on *NK* landscapes with *K* = 2 for different values of *N*. Each curve shows the fitness attained on the *y*-axis plotted against the number of tries. Each curve increases rapidly at first, and then ever more slowly, strongly suggesting an exponential slowing. (If the slowing is, in fact, exponential, reflecting the fact that at each im-

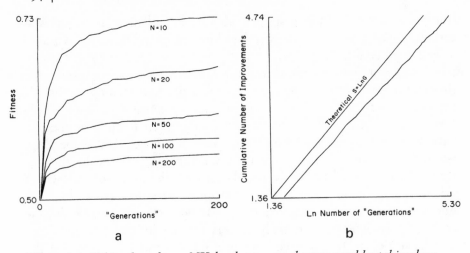

Figure 9.1 *Taking long leaps. NK landscapes can be traversed by taking long jumps—that is, mutating more than one gene at a time. But on a correlated landscape, each time a fitter long-jump variant is found, the expected number of tries to find an even better variant doubles. Fitness increases rapidly at first, and then slows and levels off. (a) This slowing is shown for a variety of K = 2 landscapes. "Generation" is the cumulative number of independent long-jump trials. Each curve is the mean of 100 walks. (b) Logarithmic scales are used to plot the number of improvements against the number of generations.*

provement the number of tries to make the next improvement really doubles, then if we replot the data from Figure 9.1*a* using the logarithm of the number of tries, we should get a linear relation. Figure 9.1*b* shows that this is true. The expected number of improvement steps, $S = \ln G$.)

The result is simple and important, and appears nearly universal. In adaptation via long jumps beyond the correlation lengths of landscapes, the number of times needed to find fitter variants doubles at each improvement step, hence slowing exponentially. It takes 1,000 tries to find the first 10 fitter variants, then 1 million tries to find the next 10, then 1 billion to find the next 10.

(Figure 9.1*a* also shows another important feature: as N increases, enlarging the space of possibilities, long-jump adaptation attains ever poorer results after the same number of tries. From other results we know that the actual heights of peaks on *NK* landscapes do not change as N increases. Thus this decrease in fitness is a further limit to selection that, in my book *The Origins of Order*, I called a complexity catastrophe. As the number of genes increases, long-jump adaptations becomes less and less fruitful; the more complex an organism, the more difficult it is to make and accumulate useful drastic changes through natural selection.)

The germane issue is this: the "universal law" governing long-jump adaptation suggests that adaptation on a correlated landscape should show three time scales—an observation that may bear on the Cambrian explosion. Suppose that we are adapting on a correlated, but rugged *NK* landscape, and begin evolving at an average fitness value. Since the initial position is of average fitness, half of all nearby variants will be better. But because of the correlation structure or shape of the landscape, those nearby variants are only *slightly* better. In contrast, consider distant variants. Because the initial point is of average fitness, again half the distant variants are fitter. But because the distant variants are far beyond the correlation length of the landscape, some of them can be *very much fitter* than the initial point. (By the same token, some distant variants can be very much worse.) Now consider an adaptive process in which some mutant variants change only a few genes, and hence search the nearby vicinity, while other variants mutate many genes, and hence search far away. Suppose that the fittest of the variants will tend to sweep through the population the fastest. Thus early in such an adaptive process, we might expect the distant variants, which are very much fitter than the nearby variants, to dominate the process. If the adapting population can branch in more than one direction, this should give rise to a branching process in which distant variants of the initial genotype, differing in many ways from one another as well, emerge rapidly. Thus early on, dramatically variant forms should arise from the initial stem. Just as in the Cambrian explosion, the species exhibiting the different major body plans, or phyla, are the first to appear.

Now the second time scale: as distant fitter variants are found, the universal law of long-jump adaptation should set in. Every time such a distant fitter variant is found, the number of mutant tries, or waiting time, to find still another distant variant doubles. The first 10 improvements may take 1,000 tries; the next 10 may take 1 million tries; the next 10 may take 1 billion tries. As this exponential slowing of the ease and rate of finding distant fitter variants occurs, then it becomes easier to find fitter variants on the local hills nearby. Why? Because the fraction of fitter nearby variants dwindles very much more slowly than in the long-jump case. In short, in the mid term of the process, the adaptive branching populations should begin to climb local hills. Again, this is what happened in the Cambrian explosion. After species with a number of major body plans sprang into existence, this radical creativity slowed and then dwindled to slight tinkering. Evolution concentrated its sights closer to home, tinkering and adding filigree to its inventions.

In the long term, the third time scale, populations may reach local peaks and stop moving or, as shown in Chapter 8, may drift along ridges of high fitness if mutation rates are high enough, or the landscape itself may deform, the locations of peaks may shift, and the organisms may follow the shifting peaks.

Recently, Bill Macready and I decided to explore the "three time scale" issue in more detail using *NK* landscapes. Bill carried out numerical studies searching at different distances across the landscape as walks proceeded uphill. Figure 9.2 shows the results.

What we wanted to know was this: as one's fitness changes, what is the "best" distance to explore to maximize the rate of improvement? Should we look a long way away, beyond the correlation length when

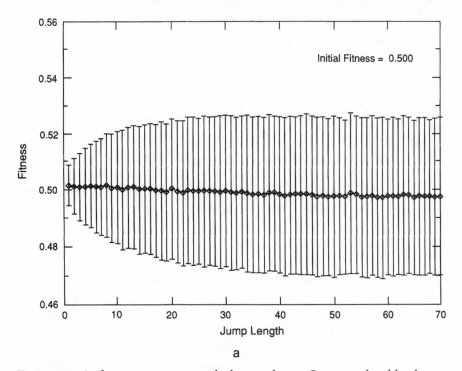

a

Figure 9.2 *As fitness increases, search closer to home. On a correlated landscape, nearby positions have similar fitnesses. Distant positions can have fitnesses very much higher and very much lower. Thus optimal search distance is high when fitness is low and decreases as fitness increases. (a) to (c) The results of sending 1,000 explorers with three different initial fitnesses to each possible search distance across the landscape. The distribution of fitnesses found by each 1,000 explorers is a bell-shaped, Gaussian curve. Crossmarks on the bars show plus or minus one standard deviation for each set of 1,000 explorers, and hence correspond to the best or worse one in six fitnesses they find.*

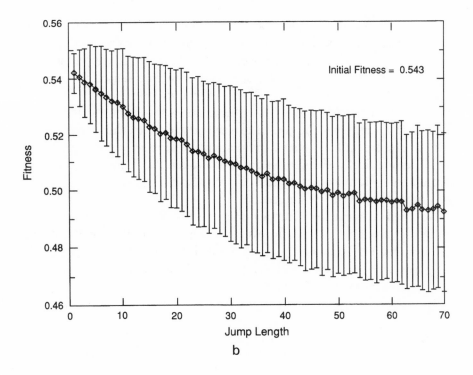

b

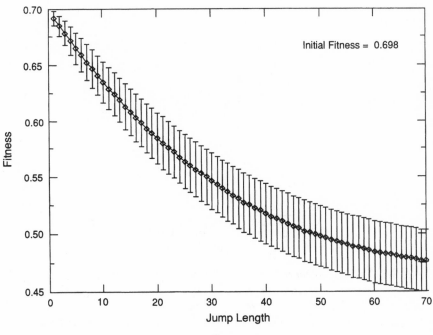

c

197

fitness is average, as I argued earlier? And, as fitness improves, should we look nearby rather than far away? As Figure 9.3 shows, summarizing the results in Figure 9.2, the answers to both questions are yes. Fitness is plotted on the *x*-axis, and the optimal distance to search to improve fitness is plotted on the *y*-axis. The implication is this: when fitness is average, the fittest variants will be found far away. As fitness improves, the fittest variants will be found closer and closer to the current position. Therefore, at early stages of an adaptive process, we would expect to find dramatically different variants emerging. Later, the fitter variants that emerge should be ever less different from the current position of the adaptive walk on the landscape.

We need to recall one further point. When fitness is low, there are many directions uphill. As fitness improves, the number of directions uphill dwindles. Thus we expect the branching process to be bushy initially, branching widely at its base, and then branching less and less profusely as fitness increases.

Uniting these two features of rugged but correlated landscapes, we

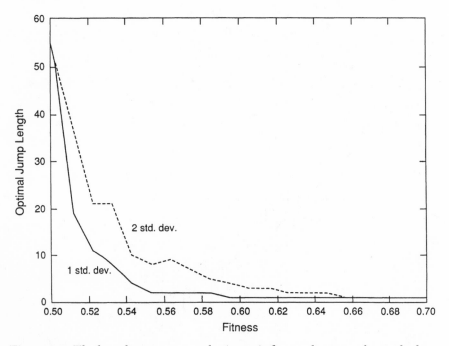

Figure 9.3 *The best distance to search. As one's fitness changes, what is the best distance to explore to maximize the rate of improvement? In this graph, as fitness increases, optimal search distance shrinks from halfway across the space to the immediate vicinity. When fitness is average, it is best to look a long way; as fitness improves, it is better to search nearby.*

should find radiation that initially both is bushy and occurs among dramatically different variants, and then quiets to scant branching among similar variants later on as fitness increases.

I believe that these features are just what we see in biological and technological evolution.

The Cambrian Explosion

From the first chapter of this book I have regaled you with images of the Cambrian explosion and the profound asymmetry of that burst of biological creativity compared with that following the later Permian extinction. In the Cambrian, over a relatively short period of time, according to most workers in the field, a vast diversity of fundamentally different morphological forms appeared. Since Linnean taxonomy has been with us, we have categorized organisms hierarchically. The highest categories, kingdoms and phyla, capture the most general features of a very large group of organisms. Thus the phylum of vertebrates—fish, fowl, and human—all have a vertebral column forming an internal skeleton. There are 32 phyla today, the same phyla that have been around since the Ordovician, the period after the Cambrian. But the best accounts of the Cambrian suggest that as many as 100 phyla may have existed then, most of which rapidly became extinct. And as we have seen, the accepted view is that during the Cambrian, the higher taxonomic groups filled in from the top down: the species that founded phyla emerged first. These radically different creatures then branched into daughter species, which were slightly more similar to one another yet distinct enough to become founders of what we now call classes. These in turn branched into daughter species, which were somewhat more similar to one another yet distinct enough to warrant classifying them as founders of orders. They in turn branched and gave off daughter species distinct enough to warrant being called founders of families, which branched to found genera. So the early pattern in the Cambrian shows explosive differences among the species that branch early in the process, and successively less dramatic variation in the successive branchings.

But in the Permian extinction some 245 million years ago, about 300 million years after the Cambrian, a very different progression unfolded. About 96 percent of all species became extinct, although members of all phyla and many lower taxa survived. In the vast rebound of diversity that followed, very many new genera and many new families were founded, as was one new order. But no new classes or phyla were formed. The higher taxa filled in from the bottom up. The puzzle is to account for the vast explosion of diversity in the Cambrian, and the profound asymmetry between the Cambrian and the Permian.

A related general phenomenon is this: during postextinction rebounds, it appears to be the case that most of the major diversification occurs early in the branching process of speciation. Paleontologists call such a branching lineage a *clade*. They speak of "bottom-heavy" clades, which are bushy at the base, or oldest time, and note that genera typically diverge early in the history of their families, while families diverge early in the history of their orders. In short, the record seems to indicate that during postextinction rebounds most of the diversity arises rather rapidly, and then slows. Thus while the Cambrian filled in from the top down and the Permian from the bottom up, in both cases the greatest diversification came first, followed by more conservative experimentation.

Might it be the case that the general features of rugged fitness landscapes shed light on these apparent features of the past 550 million years of evolution? As I have suggested, the probable existence of three time scales in adaptive evolution on correlated rugged landscapes, summarized in Figure 9.3, sounds a lot like the Cambrian explosion. Early on in the branching process, we find a variety of long-jump mutations that differ from the stem and from one another quite dramatically. These species have sufficient morphological differences to be categorized as founders of distinct phyla. These founders also branch, but do so via slightly closer long-jump variants, yielding branches from each founder of a phylum to dissimilar daughter species, the founders of classes. As the process continues, fitter variants are found in progressively more nearby neighborhoods, so founders of orders, families, and genera emerge in succession.

But why, then, was the flowering after the Permian extinction so different from the explosion during the Cambrian? Can our understanding of landscapes afford any possible insight? Perhaps. A few more biological ideas are needed. Biologists think of development from the fertilized egg to the adult as a process somewhat akin to building a cathedral. If one gets the foundations wrong, everything else will be a mess. Thus there is a common, and probably correct, view that mutants affecting early stages of development disrupt development more than do mutants affecting late stages of development. A mutation disrupting formation of the spinal column and cord is more likely to be lethal than one affecting the number of fingers that form. Suppose this common view is correct. Another way of saying this is that mutants affecting early development are adapting on a more rugged landscape than mutants affecting late development. If so, then the fraction of fitter neighbors dwindles faster for mutants affecting early development than those affecting late development. Thus it becomes hard to find mutants altering early development sooner in the evolutionary process than to find mutants affecting late development. Hence if this is correct, early development tends to "lock in" before late development. But alterations in early development are just the ones that would cause sufficient morphological change to count as change at the phylum or class level. Thus as the evolutionary process continues and early development locks in, the most rapid response to ecological opportunity after a mass-extinction event should be a rebound with massive speciation and radiation, but the mutations should affect late development. If this is true, no new phyla or classes will be found. The radiation that occurs will be at the genus and family level, corresponding to minor changes that result from mutations affecting late development. Then the higher taxa should fill in from the bottom up.

In short, if we imagine that by the Permian early development in the organisms of most phyla and classes was well locked in, then after 96 percent go extinct, only traits that were more minor, presumably those caused by mutations affecting later stages of an organism's ontogeny, could be found and improved rapidly.

If these views are correct, then major features of the record, including wide radiation that fills taxa from the top down in the Cambrian, and the asymmetry seen in the Permian, may find natural explanations as simple consequences of the structure of fitness landscapes. In the same vein, notice that bushy radiation should generally yield the greatest morphological variation early in the process. Thus one might expect that during postextinction rebounds, genera would arise early in the history of their families and families would arise early in the history of their orders. Such bottom-heavy clades are just what is observed repeatedly in the evolutionary record.

Technological Evolution on Rugged Landscapes

At first glance, the adaptive evolution of organisms and the evolution of human artifacts seem entirely different. After all, Bishop Paley urged us to envision a watchmaker to make watches and God the watchmaker to make organisms, and then Darwin pressed home his vision of a "blind watchmaker" in his theory of random variation and natural selection. Mutations, biologists believe, are random with respect to their prospective significance. Man the toolmaker struggles to invent and improve, from the first unifacial stone tools some 2 or more million years ago, to the bifacial hand axes of the lower Paleolithic, to the superbly crafted flint blades hammered free from prepared cores and then pressure-flaked to stunning perfection. What in the world can the blind process of adaptive evolution in biological organisms have to do with technological evolution? Perhaps nothing, perhaps a great deal.

Despite the fact that human crafting of artifacts is guided by intent and intelligence, both processes often confront problems of conflicting constraints. Furthermore, if Darwin proposed a blind watchmaker who tinkered without foreknowledge of the prospective significance of each mutation, I suspect that much of technological evolution results from tinkering with little real understanding ahead of time of the consequences. We think; biological evolution does not. But when problems are very hard, thinking may not help that much. We may all be relatively blind watchmakers.

Familiar features of technological evolution appear to bespeak search on rugged landscapes. Indeed, qualitative features of technological evolution appear rather strikingly like the Cambrian explosion: branching radiation to create diverse forms is bushy at the base; then the rate of branching dwindles, extinction sets in, and a few final, major alternative forms, such as final phyla, persist. Further, the early diversity of forms appears to be more radical, and then dwindles to minor tuning of knobs and whistles. The "taxa" fill in from the top down. That is, given a fundamental innovation—gun, bicycle, car, airplane—it appears to be common to find a wide range of dramatic early experimentation with radically different forms, which branch further and then settle down to a few dominant lineages. I have already mentioned, in Chapter 1, the diversity of early bicycles in the nineteenth century: some with no handlebars, then forms with little back wheels and big front wheels, or equal-size wheels, or more than two wheels in a line, the early dominant Pennyfarthing branching further. This plethora of the class Bicycle (members of the phylum Wheeled Wonders) eventually settled to the two or three forms dominant today: street, racing, and mountain bike. Or think of the highly diverse forms of steam and gasoline flivvers early in the twentieth century as the automobile took form. Or of early aircraft design, helicopter design, or motorcycle design. These qualitative impressions are no substitute for a detailed study; however, a number of my economist colleagues tell me that the known data show this pattern again and again. After a fundamental innovation is made, people experiment with radical modifications of that innovation to find ways to improve it. As better designs are found, it becomes progressively harder to find further improvements, so variations become progressively more modest. Insofar as this is true, it is obviously reminiscent of the claims for the Cambrian explosion, where the higher taxa filled in from the top down. Both may reflect generic features of branching adaptation on rugged, correlated fitness landscapes.

Learning Curves

A second signature that technological evolution occurs on rugged fitness landscapes concerns "learning curves" along technological trajectories. There are two senses in which this occurs. First, the more copies of an item produced by a given factory, the more efficient production becomes. The general result, as accepted by most economists, is this: at each doubling of the number of units produced in a factory, the cost per unit (in inflation-adjusted dollars or in labor hours) falls by a constant fraction, often about 20 percent. Second, learning curves also arise on what are called technological trajectories. It appears common that the rate of improvement of various technologies slows with total industry expenditure; that is, improvement in performance is rapid at first, and then slows.

Such learning curves show a special property called a power-law relation. A simple example of such a power law would be this: the cost in labor hours of the Nth unit produced is $1/N$ of the cost of the first unit produced. So if you make 100 widgets, the last one costs only 1/100 as much as the first. The special character of a power law shows up when the logarithm of the cost per unit is plotted against the logarithm of the total number of units produced. The result is a straight line showing the cost per unit decreasing as the total number of units, N, increases.

Economists are well aware of the significance of learning curves. So too are companies, which take them into account in their decisions on budgets for production runs, projected sales price per unit, and the expected number of units that must be sold before a profit is made. In fact, the power-law shapes of these learning curves are of basic importance to economic growth in the technological sector of the economy: during the initial phase of rapid improvements, investment in the new technology yields rapid improvement in performance. This can yield what economists call increasing returns, which attract investment and drive further innovation. Later, when learning slows, little improvement occurs per investment dollar, and the mature technology is in a period of what economists call diminishing returns. Attracting capital for further innovation becomes more difficult. Growth of that technology sector slows, markets saturate, and further growth awaits a burst of fundamental innovation in some other sector.

Despite the ubiquity and importance of these well-known features of technological evolution and economic growth, no underlying theory seems to account for the existence of learning curves. Do our simple insights into adaptive processes on landscapes offer any help? Again, perhaps, and the story is a typical Santa Fe Institute adventure. In 1987, John Reed (chairman of Citicorp) asked Phil Anderson (a Nobel laureate in physics) and Ken Arrow (a Nobel laureate in economics) to organize a meeting to bring economists together with physicists, biologists, and others. The institute had its first meeting on economics and established an economics program, first headed by the Stanford economist Brian Arthur. I, in turn, began trying to apply ideas about fitness landscapes to technological evolution. Several years later, two young economics graduate students, Phil Auerswald of the University of Washington and José Lobo from Cornell, were taking the institute's summer course on complexity, and asked if they might work with me on applying these new ideas about landscapes to economics. José began talking with the Cornell economist Karl Shell, already a friend of the institute. By the summer of 1994, all four of us began collaborating, helped by Bill Macready, a solid-state physicist and postdoctoral fellow working with me at the institute, and Thanos Siapas, a straight-A computer-science graduate student at MIT. Our preliminary results suggest that the now familiar NK model may actually account for a number of well-known features of learning curves: the power-law relationship between cost per unit and total number of units produced; the fact that after increasingly long periods with no improvement, sudden improvements often occur; and the fact that improvement typically reaches a plateau and ceases.

Recall that on random landscapes, every time a step is taken uphill, the number of directions uphill falls by a constant fraction, one-half. More generally, we saw that with the NK landscape model for K larger than perhaps 8, the number of fitter neighbors dwindles by a constant fraction at each step toward higher fitness. Conversely, the number of "tries" to find an improvement increases by a constant fraction after each improvement is found. Thus the rate of finding fitter variants—of making incremental improvements—shows exponential slowing. The particular rate of exponential slowing depends, in the NK model, on K. The slowing is faster when the conflicting constraints, K, are higher and the landscape is more rugged. Finally, recall that adaptive walks on rugged landscapes eventually reach a local optimum, and then cease further improvement.

There is something very familiar about this in the context of technological trajectories and learning effects: the rate of finding fitter variants (that is, making better products or producing them more cheaply) slows exponentially, and then ceases when a local optimum is found. This is already almost a restatement of two of the well-known aspects of learning effects. First, the total number of "tries" between finding fitter variants increases exponentially; thus we expect that increasingly long periods will pass with no improvements at all, and then rapid improvements as a fitter variant is suddenly found. Second, adaptive walks that are restricted to search the local neighborhood ultimately terminate on local optima. Further improvement ceases.

But does the *NK* model yield the observed power-law relationship? To my delight, the answer appears to be yes. We already know that the rate of finding fitter variants slows exponentially. But how much improvement occurs at each step? In the *NK* model, if the "fitness" values are considered instead as "energy" or "cost per unit," and adaptive walks seek to minimize energy or "cost," then it turns out that with each of these improvements the reduction in cost per unit is roughly a constant fraction of the improvement in cost per unit achieved the last time an improvement was made. Thus the amount of cost reduction achieved with each step slows exponentially, while the rate of finding such improvements also slows exponentially. The result, happy for us four, is that cost per unit decreases as a power-law function of the total number of tries, or units produced. So if the logarithm of cost per unit is plotted on the *y*-axis, and the logarithm of the total number of tries, or units produced, is plotted on the *x*-axis, we get our hoped-for straight-line (or near straight-line) distribution.

Not only that, but to our surprise—and, at this stage of our work, healthy skepticism—not only does a power law seem to fall out of the good old *NK* model, but we find power laws with about the right slopes to fit actual learning curves.

You should not take these results as proof that the *NK* model itself is a proper microscopic account of technological evolution. The *NK* model is merely a toy world to tune our intuitions. Rather, the rough successes of this first landscape model suggests that better understanding of technological landscapes may yield deeper understanding of technological evolution.

I am not an expert on technological evolution; indeed, I am also not an expert on the Cambrian explosion. But the parallels are striking, and it seems worthwhile to consider seriously the possibility that the patterns of branching radiation in biological and technological evolution are governed by similar general laws. Not so surprising, this, for all these forms of adaptive evolution are exploring vast spaces of possibilities on more or less rugged "fitness" or "cost" landscapes. If the structures of such landscapes are broadly similar, the branching adaptive processes on them should also be similar.

Tissues and terra-cotta may indeed evolve in similar ways. General laws may govern the evolution of complex entities, whether they are works of nature or works of man.

Chapter 10

An Hour upon
the Stage

D arwin's own image of it is the tangled bank, an ecosystem rife with
the ebullience of life—hawthorn, ivy, earthworms, finches, spar-
rows, moths, pillbugs, untold kinds of beetles, let alone squirrels, foxes,
frogs, ferns, lilacs, elderberry, and soft mosses. A century later, Dylan
Thomas sang of his native Wales:

> On a breakneck of rocks
> Tangled with chirrup and fruit,
> Froth, flute, fin and quill
> At a wood's dancing hoof.

Tangled and interwoven, dancing together in rhythms, cadences, and
profusion. The miracle is so wondrous, the more so because there is no
choreographer. Each organism lives in the niche created by the artful
ways of other organisms. Each seeking its own living blindly creates the
modes of living for the others. An ecosystem is a tangled bank of interwoven roles—metabolic, morphological, behavioral—that is magically
self-sustaining. Sunlight is gathered; carbon dioxide and water are built
up to sugars; nitrogen is fixed and built into amino acids—the captured
energy driving linked metabolisms within cells, within organisms, and
among organisms.

Four billion years ago, some congery of molecules danced, blindly
catalyzing one another's formation, reaching a critical diversity at which
self-sustaining webs of reactions emerged and formed life. Blind interactions gave rise to the emergent phenomena of cellular life, cells linked
in metabolic exchanges blindly creating the first ecosystems. Those
ecosystems have unfolded for over billions of years with a profusion of
species emerging and passing into extinction. And at each level, we
sense emergent lawfulness in the profusion.

David Raup, a superb paleontologist at the University of Chicago, estimates that between 99 percent and 99.9 percent of all species that have ever existed are now extinct. The earth today may harbor 10 million to 100 million species. If so, then life's history may have seen 10 billion to 100 billion species come and go. One hundred billion players strutting and fretting their hour upon the stage, and then are heard no more.

We know so little about the means by which open thermodynamic systems, like the earth, create order. But many of us sense lawfulness in the profusion. For all our ignorance, there are hints of law at three levels. The first level is that of a community or an ecosystem, in which species assemble and make their livings in the niches each provides the others. The second level, often occurring on a longer time scale than community assembly and ecological change, is that of coevolution. Species not only evolve on their fitness landscapes, as we discussed in Chapters 8 and 9, but coevolve with one another. The idealization we have used that fitness landscapes are fixed and unchanging is false. Fitness landscapes change because the environment changes. And the fitness landscape of one species changes because the other species that form its niche themselves adapt on their own fitness landscapes. Bat and frog, preditor and prey, coevolve. Each adaptive move by the bats deforms the landscape of the frogs. Species coevolve on coupled dancing landscapes.

But there is yet a third, still higher level, presumably occurring on a still longer time scale than coevolutionary processes. The coevolution of organisms alters both the organisms themselves and the ways organisms interact. Over time, the ruggedness of the fitness landscapes changes, as does their resiliency—how easily each landscape is deformed by the adaptive moves of the players. The very process of coevolution itself evolves!

Nowhere in these rhythms is there a master choreographer. Selection acts at the level of the individual organism. Selection picks out the fitter variant *individuals*—those likely to leave the most offspring. Selection, biologists think, almost certainly does not act at the level of groups of individuals, selecting the fittest among competing groups. Nor does selection act at the level of whole species or whole ecosystems. The vast puzzle is that the emergent order in communities—in community assembly itself, in coevolution, and in the evolution of coevolution—almost certainly reflects selection acting at the level of the individual organism. Adam Smith first told us the idea of an invisible hand in his treatise *The Wealth of Nations*. Each economic actor, acting for its own selfish end, would blindly bring about the benefit of all. If selection acts only at the level of the individual, naturally sifting for fitter variants that "selfishly" leave more offspring, then the emergent order of communities, ecosystems, and coevolving systems, and the evolution of coevolution itself are the work of an invisible choreographer. We seek the laws that constitute that choreographer. And we will find hints of such laws, for the evolution of coevolution may bring coevolving species to hover ever poised between order and chaos, in the region I have called the edge of chaos.

Community

Ecologists who think about community dynamics and community assembly have a variety of well-understood theories to draw on. The first concerns population dynamics in ecosystems with predators and prey, or other webs of interaction. In the early decades of this century, two theoretical biologists, A. J. Lotka and V. J. Volterra, established much of the conceptual foundation still in use and formulated simple models of the increases and decreases in abundance of different species in a community as the species interact.

Consider a hypothetical ecosystem with grass, rabbits, and foxes—hence with plants, herbivores, and carnivores, one of each. The simplest models suppose that the grass grows and, in the simplest case, is present at a constant amount per square mile. The rabbits eat the grass, mate, and do what rabbits do to give birth to baby rabbits. The foxes hunt for rabbits, eat the rabbits, mate, and have baby foxes. The theory develops by writing down an equation for the rate of increase or decrease of the rabbit population as a function of the current rabbit and fox populations, and writing down a separate equation for the rate of increase or decrease of the fox population as a function of the current rabbit and fox populations. Each is a "differential equation," which just means an equation expressing the rate of change of something—for example, the change in the number of members of a population—as a function of something else—in this case, the present number of foxes and rabbits. "Solving" the equations just means starting the model ecosystem with a particular population of foxes and rabbits per square mile, and following the "predictions" of the equations showing how these populations increase or decrease in time.

Commonly in such models, the community either settles to a steady state or, alternatively, shows persistent oscillations. Figure 10.1 shows these two kinds of behavior, plotting the fox population and the rabbit population on the *y*-axis, and the time elapsed from the start on the *x*-

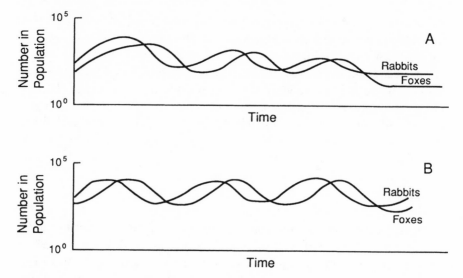

Figure 10.1 *A hypothetical ecosystem with foxes and rabbits. The size of the current population is plotted against the passage of time. (a) The system settles to a steady state at which each population remains constant. (b) The populations of foxes and rabbits settle into a sustained oscillation, called a limit cycle.*

axis. In the first instance, the initial rabbit and fox populations may change for a while, perhaps increasing or decreasing, but eventually settle down to unchanging levels. In the second case, the initial rabbit and fox populations enter into a pattern of increase and decrease—a persistent oscillation called a limit cycle. At first, the fox population is low, so foxes are not around to eat the rabbits as fast as rabbits breed to make more rabbits. Hence, at first, the rabbit population increases. Soon there are so many rabbits to feed the foxes that the fox population starts to grow. But as the fox population grows, the foxes eat rabbits faster than rabbits reproduce. Hence the rabbit population falls. When the rabbit population falls, there are not enough rabbits to feed the foxes, so the fox population dies off. But then the fox population is low, so the rabbits, free of the foxes, start to increase in abundance again. Thus the oscillation continues indefinitely.

Such oscillations are well known in real ecosystems. Volterra was led to his role in the Lotka–Volterra equations by reports of oscillations in the commercial fish catch in the Adriatic. Sustained oscillations of arctic fox and hare have been recorded over many cycles. Even more complex behaviors, such as chaos, with its fabled butterfly, can arise in model and probably in real ecosystems. In fact, much of the early work on chaos theory was carried out by mathematical ecologists. In these systems, very small changes in initial conditions—the number of rabbits or foxes, for example—cause the future evolution of the system to change dramatically.

Models like those of Lotka and Volterra have provided ecologists with simple "laws" that may govern predator–prey relationships. Similar models study the population changes, or population dynamics, when species are linked into more complex communities with tens, hundreds, or thousands of species. Some of these links are "food webs," which show which species eat which species. But communities are more complex than food webs, for two species may be mutualists, may be competitors, may be host and parasite, or may be coupled by a variety of other linkages. In general, the diverse populations in such model communities might exhibit simple steady-state patterns of behavior, complex oscillations, or chaotic behavior.

As one might expect, the populations of some species can sometimes decrease to the point of vanishing. In the meantime, other species might migrate into the community, altering the web of interactions and the population dynamics. One species, by entering a community, might drive others to extinction. Extinction of one species might increase the abundance of other species or drive still other species to extinction.

These facts lead to the next level of ecological theory: community assembly. The basic question here is: How do stable communities of species come together? The answers are not known. Experimental work has looked at a variety of issues. For example, suppose one takes a patch of prairie, or a patch of Sonoran desert, and puts a fence around it so that certain types of small animals can no longer enter the plot. The kinds of plants in the plot change over time. Now one might think that if the fence were removed, allowing the same set of small animals back in, one would get back the original community of plants. It appears that this intuition is wrong. Typically, one gets a *different* stable community! It seems that given a "reservoir" of species that can migrate into a plot or patch, the community that forms is deeply dependent on the sequence in which species are introduced. My friend, ecologist Stuart Pimm, has coined the term "Humpty Dumpty effect": you cannot always put an ecosystem back together with only the final species in the community. Pimm points to a nice example: how North American prairie plant and animal communities changed from one interglacial period to the next. In the past 10,000 years, there have been humans, bison, and antelope. During the previous interglacial period, there were very many more species of large mammals—horses, camels, ground sloths, and others. And the plant communities changed, too. The communities involved different mixes of species each time.

Pimm and his colleagues have struggled to understand these phenomena and have arrived at ideas deeply similar to the models of fitness landscapes we discussed in Chapters 8 and 9. Different communities are imagined as points on a *community landscape*. Change the initial set of species, and the community will climb to a different peak, a different stable community. They model community assembly using Lotka–Volterra-type equations. Pimm and his colleagues imagine a hypothetical reservoir of species: some, like grass, just grow; others, like rabbits, eat grass and grow; others, like foxes, eat rabbits and grow.

In these models, Pimm and friends toss randomly chosen species into a "plot" and watch the population trajectories. If any species goes to zero population, hence extinct, it is "removed" from the plot. The results are both fascinating and still poorly understood. What one finds is that, at first, it is easy to add new species, but as more species are introduced, it becomes harder and harder. That is, more randomly chosen species must be tossed into the plot to find one that can survive with the rest of the assembling community. Eventually, the model community is *saturated* and stable; no further species can be added. But if the simulation is run again and again, with the same reservoir of hypothetical species, the resulting stable communities can be very different, depending on the order in which species are introduced. Further, one can ask what happens to a stable community if one of the species is removed. One finds that the community may experience avalanches of extinction events. Such an avalanche of extinctions occurs because chain reactions arise. Extinction of a specific kind of grass leads to extinction of some herbivore that eats that grass; then, in turn, some of the carnivores that eat the now absent herbivores are driven extinct. Conversely, removal of a carnivore can decrease the predation on the two herbivore species and can allow one to outcompete the other for grass, driving the second one extinct. In these studies, one finds that many small avalanches of extinctions occur, and few large ones.

The results of Pimm and his colleagues may be very important and very general. They seem to be finding the special kind of distribution of avalanches of extinction events called a power-law distribution. Suppose we plot the size of an extinction event, measured as the number of species that went extinct, on the x-axis, and we plot the number of extinction events of a given size on the y-axis (Figure 10.2a). There are many small extinction avalanches and few large ones. In a true power-law relation, the number of avalanches of a given size should fall as the size of the avalanche raised to some power. There is a simple way to test for this. We plot the logarithm of the size of the extinction avalanche on the x-axis, and the logarithm of the number of extinction events of a

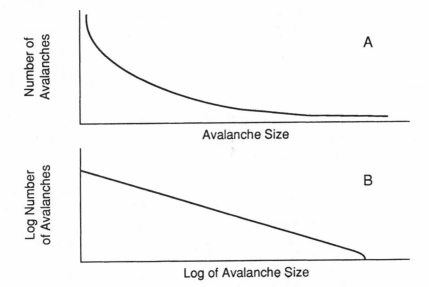

Figure 10.2 *Avalanches of extinction. A hypothetical distribution of avalanches of extinction events.* (a) *The size (number of species involved) in an extinction event is plotted against the number of avalanches of each size, showing that there are more smaller avalanches and fewer large ones.* (b) *The same hypothetical data is replotted using logarithmic scales. The result is a linear relation, the hallmark of a power-law distribution.*

given size on the y-axis. If we plot a power-law relationship using logarithmic scales, a so-called log–log plot, we get a straight line—the acid test that such a relationship really exists (Figure 10.2*b*).

Two features of such distributions are of the greatest interest. First, a power-law means that extinctions can arise on all size scales. The only limit to the size of the largest avalanche is the number of species in the entire system. Small extinction avalanches are common, but if one waits long enough, eventually an extinction event of any arbitrary size will be encountered. The second property of deep interest is that the same small initial event, removal of a single species in this case, can trigger both small and vast extinction events. Systems that show power-law behavior are often "poised" so the very same small event might trigger a small or a catastrophic avalanche of change.

We keep finding hints of power-law distributions at a variety of levels. In Chapters 4 and 5, we examined the "edge of chaos" in Boolean genomic networks. Recall that alteration in the activity of a single on–off gene could trigger a cascade of alterations. At the phase-transition region between order and chaos, the size distributions of such cascades or avalanches is almost certainly a power-law distribution, with many small and fewer large avalanches sending signals of change propagating across the system. Indeed, power-law distributions arise at many phase transitions. In thinking about the metabolic diversity of ecosystems in Chapter 6, we were led to the possibility that ecosystems may evolve to the subcritical–supracritical boundary. There, the "branching probability" of chain reactions forming new molecules is exactly 1.0. This should also yield a power-law distribution of bursts of molecular novelty. Now we find hints of power-law distributions of extinction events in models of community assembly. Later in this chapter, we will find power-law distributions of extinction events in other models of co-evolution. And we will find fossil evidence of approximate power laws in actual species-extinction events. These are all models of far-from-equilibrium systems; all are exhibiting similar emergent regularities. Perhaps a general law is awaiting elucidation.

The community-assembly simulation studies are fascinating for a number of reasons beyond the distribution of extinction events. In particular, it is not obvious why model communities should "saturate," so that it becomes increasingly difficult and finally impossible to add new species. If one constructs a "community landscape," in which each point of the terrain represents a different combination of species, then the peaks will represent points of high fitness—combinations that are stable. While a species navigates a fitness landscape by mutating genes, a community navigates a community landscape by adding or deleting a species. Pimm argues that as communities climb higher and higher toward some fitness peak, the ascension becomes harder and harder. As the climb proceeds, there are fewer directions uphill, and hence it is harder to add new species. At a peak, no new species can be added. Saturation is attained. And from one initial point the community can climb to different local peaks, each representing a different stable community.

Pimm concedes that the use of the landscape metaphor here is worrisome. The trouble is that he is visualizing a space of all the possible communities that might be assembled from his hypothetical reservoir of species. Community assembly then consists in passing from a community to a neighboring community by adding or deleting a species. Now, if there were something called "the fitness of the community," one would legitimately get our old friend, the fitness landscape. The trouble is that it is not obviously sensible to talk about a "community" fitness. Whether I, the Eastern grama grass, or squirrels invade a patch has to do with whether it is good for me, whether I can make it in this new environment with the other species. My success in invading is not obviously dependent on whether my doing so creates a "fitter community."

Yet Pimm's simulations do actually behave somewhat *as though* there were such a "community fitness," as though communities were indeed climbing to fitness peaks.

So at least in the model, if not in life, we see an emergent phenomenon. The decisions about how species interact—who preys on whom, who is parasite and who is host, and so forth—are drawn from some random distribution. And so, without Stuart Pimm or his colleagues acting as intentional choreographers, without even understanding why it occurs, the model communities behave as though they were climbing uphill to attain local peaks, stable communities. An invisible hand seems to be at work.

Coevolution

In the simplified models considered so far, the species themselves are fixed; they do not evolve. If we want to construct models that will help us understand real ecosystems, we will have to consider what happens with species that change as they interact with one another, species that coevolve.

Species live in the niches afforded by other species. They always have and presumably always will. Once initial life arose and began to diversify, exchanging molecules that might poison or feed one another, organisms joined into a coevolutionary dance, jockeying for places next to one another as mutualists, competitors, predators and prey, or hosts and parasites.

Flowers coevolved with the insects that pollinated them and fed on their nectar. Here was a mutualism over thousands of millennia, breeding the beauty of an early autumn field with bees still searching, signalling the discovery of the riches with their dance. The root nodule seeps carbohydrates to bacteria that fix nitrogen for the plant's further uses. We seep carbon dioxide to the plants, which seep oxygen to us. We all swap our stuff. Life is a vast Monopoly game, with energy the ultimate currency, and the sun our banker of last resort.

Such coevolving mutualism is more common than previously thought, and it can be intensely intimate. As we saw, it now appears that the complex eukaryotic cells that form multicellular organisms evolved by forming an endosymbiotic alliance in which formerly free bacteria became mitochondria and chloroplasts. Within the mitochondria one still finds an autonomous genomic system, thought to be the remains, at least partially, of the genomic system in the free-living bacterial ancestor. Imagine the intricacy of linking the metabolism of two cells, host and endosymbiont, so that each benefits the other, the mitochondria dividing at a rate that maintains a stable population within each cell, while the cell enjoys the energetic fruits of the effort. Bacteria have their own endosymbionts of a sort, circles of DNA, called plasmids, that can divide only in bacterial host cells, and carry with them molecular tricks such as resistance to certain antibiotics.

Coevolution extends beyond mutualism and symbiosis. Host–parasite systems coevolve, from malaria to the AIDS agent, HIV. The agent of malaria alters its surface antigens to evade detection by the host; the host immune system evolves to try to match, catch, and destroy the malaria. A molecular dance of hide-and-seek takes place. The same coevolutionary dance, dramatic and often tragically lethal, now takes place in those infected with the HIV virus. Rapid mutational alteration takes place in the growing population of viruses within one HIV-positive person. We know this by detailed DNA sequencing of the virus population from a single individual. As with the organisms in the Cambrian explosion, or in postextinction rebounds, rapid radiation of viral sequences appears to occur. One theory suggests that this diversification is driven by the evolution of the viruses to evade the immune response mounted to the HIV infection. The antibodies are also evolving as the immune system tries to match the virus. Again, molecular hide-and-seek. The human immune system and HIV coevolve. Tragically, HIV currently has a deadly upper hand, for the cells it happens to invade, rather than being the mucosa of the larynx and yielding a laryngitis, are the helper T-cells of the immune system itself. Attempts to find rational therapies are, in part, based on trying to block HIV attachment and entry into helper T-cells. Unfortunately, HIV appears to evolve within the host too fast to catch easily.

Coevolution also occurs between predator and prey species. The evolution of calcified fortifications such as spurs on certain shells has, presumably, been driven by the capacity of starfish to capture and open less complex shell forms. Starfish have responded in kind with sharper, stronger beaks, larger size, and stronger suction to grip their prey. This form of persistent coevolution has been called an "arms race" or the "Red Queen Effect." Lee Van Valen, a paleontologist at the University of Chicago, coined the latter term based on the comment by the Red Queen to Alice: "[I]t takes all the running you can do, to keep in the same place." In a coevolutionary arms race, when the Red Queen dominates, all species keep changing and changing their genotypes indefinitely in a never-ending race merely to sustain their fitness level.

Coevolution appears to be a powerful aspect of biological evolution. Establishing that any pair of species or organisms within a species undergo coevolution is not trivial, and some evolutionary biologists doubt how general and powerful a process coevolution is. But most biologists feel that the genetic dance of organisms with one another is one of the major features of biological evolution.

Coevolution, in a generalized sense, is occurring in our economic and cultural systems as well. The goods and services in an economic web exist only because each is useful in either the fabrication or the creation of another good or service, as an intermediate good, or is useful to some ultimate consumer. Goods and services "live" in niches created by other goods and services. The mutualism of the biosphere, where advantages of trade exist, finds its mirror in economic systems, where advantages of trade exist in the vast web of goods and services. I will pursue this parallel later. It seems clear that at least an analogy exists between the unrolling panorama of interacting, coevolving species—each coming into existence and making a living in the niche afforded by the other species, each becoming extinct in due course—and the way that technological evolution drives the speciation and extinction of technologies, goods, and services. We are all hustling our wares—bacteria, fox, CEO. Moreover, we are all creating niches for one another. I suspect that there is more than an analogy. I suspect that biological coevolution and technological coevolution, the increasing diversity of the biosphere and of our "technosphere," may be governed by the same or similar fundamental laws.

How do biologists think about coevolution? The dominant framework, a good one I think, is based on *game theory.* Game theory was invented by the mathematician John von Neumann and developed, together with the economist Oskar Morgenstern, to think about rational economic agents. So before we spin off briefly into the world of game theory, it is well to emphasize a critical distinction. While economic agents, you and I, might have foresight and plan ahead about the consequences of our action, applications of game theory in evolution have to be consistent with our fundamental assumption that mutations occur at random with respect to prospective effects on fitness. I might have rational plans for my selfish benefit; an evolving population of bacteria does not. Natural selection may pick the fitter variants, but the bacteria harboring the mutation did not "try" to mutate in a particular way for its selfish benefit.

The simplest idea of a game is exemplified by the famous "Prisoner's Dilemma." You and I have been caught by the police. We are each being grilled in a separate cell. The cops tell me that if I rat on you and you have not ratted on me, then I go free. They tell you the same thing. If you squeal and I have kept the silence, you skip out. In either of these cases, the honest crook, whoever has kept the silence, hits the slammer for 20 years. If we both rat on each other, we each get stiff sentences, but not as stiff as if one rats and one stays mum. Say we both get 12 years. If we both shut up and stay silent, we both get lighter sentences, 4 years.

Well, you see the dilemma. Call not talking "cooperating," and call telling the police "defecting." One natural behavior is that we both wind up defecting and ratting on each other. I figure I am better off if I squeal, since if you don't, I get out free. And even if you do squeal, I'll serve less time than if I remain silent and you defect. You figure the same thing. We both defect and wind up in jail for 12 years.

The underlying issue in game theory is this: a game consists in a set of payoffs to each of a set of players. Each player has a set of "strategies" to choose from. The payoff from each strategy depends on the strategies chosen by the other players. If each player acts for his or her own selfish advantage, what kinds of *coordinated actions* will emerge? Game theory attempts to look in precise ways at invisible hands coordinating the action of independent agents.

A remarkable theorem by John Nash, an early pioneer in game theory, shows that there is always at least one "Nash" strategy for each player, with the property that if each player chooses that strategy, he or she will be better off than with any other strategy—as long as all the other players choose their own Nash strategies. Such as set of strategies among the players is called a Nash equilibrium. The defect–defect strategy in Prisoner's Dilemma is an example: if you are going to defect, then I am worse off if I do anything but defect. Since the same is true for you, defect–defect is a Nash equilibrium. The concept of Nash equilibria was remarkable insight, for it offers an account of how independent, selfish agents might coordinate their behavior without a master choreographer.

While Nash equilibria are fascinating, this concept of a "solution" of a game has major defects. The Nash equilibrium may not be one in which the payoff to the agents is particularly good. Further, in large games with many players, each having many alternative strategies, there may be many Nash equilibria. None of them may have high payoffs compared with some other sequence of actions, and there is no guaranteed way the players can choose the best among the Nash equilibria or even "find" these Nash strategies in the large spaces of possibilities.

It is precisely the fact that Nash equilibria may not have the best possible payoff that has driven the interest in the Prisoner's Dilemma. The Nash equilibrium, in which both players defect and get 12 years in jail, is far worse for both players than the cooperate–cooperate solution, in which both play mum and do 4 years in jail. Unfortunately, the highest payoff strategy, cooperate–cooperate, is not stable to defection, for if you talk and I do not, you get off free. So if I opt to stay silent, hence cooperate, I run the risk that you will defect. I will go to jail for 20 years; you will get off scot-free. You run the same risk.

Prisoner's Dilemma has been studied a great deal to understand the conditions under which cooperation will emerge. The rough answer is that if you and I play a repeated game and we do not know how many times we will play the same game, then cooperation does tend to emerge. For a single game, the rational strategy is defect–defect, but surprisingly in a repeated game, entirely different strategies emerge. To explore this, Robert Axelrod, a political scientist at the University of Michigan, MacArthur Fellow, and colleague at the Santa Fe Institute, ran a tournament in which populations of programs played repeated games of Prisoner's Dilemma with one another. The results have been of wide interest. Among the best "strategies" to emerge is "tit for tat." Here each player cooperates unless the other defects. In that case, the first player defects in the next round, tit for tat, and then resumes cooperating. Tit for tat is a stable strategy in the sense that it is better than many other strategies and pays off well if all players use it. Since the number of possible strategies in repeated Prisoner's Dilemma is enormous—one tit for two tats, always defect, always cooperate, or other complex patterns of response—it is not known if tit for tat is the best possible strategy in repeated Prisoner's Dilemma. But it is intriguing that some benign cooperation among selfish agents emerges despite the constant temptation to defect.

John Maynard Smith, an outstanding evolutionary biologist and an old friend, came to the University of Chicago in 1971 when I was a young faculty member and a recent M.D. John's main aim was to figure out how to apply game theory to evolution. His visit was a delight, for he is a master of the British tradition of "tea." As we sat down for this afternoon ritual, John would describe his efforts at studying community models and population dynamics. At the time, there was a fundamental flaw in his computer simulations: the number of rabbits kept falling well below zero. It is very unfortunate if your theory predicts the presence of negative rabbits. John must have been deeply upset, for he contracted a minor pneumonia, which I was able to diagnose at the emergency room where I was working. Cured it, I did. John, in return, taught me to "do sums," as he called it, about my Boolean nets. With his help, I was able to prove some early theorems about the emergence of the frozen components in Boolean networks described in Chapters 4 and 5.

John struggled to formulate a version of game theory for evolutionary biology by generalizing the idea of a Nash equilibrium to that of an evolutionary stable strategy (ESS). We have already described the idea of a genotype space in Chapter 8: each sequence of genes is represented by a point in a high-dimensional space of possible genotypes. Now think of each genotype as a "strategy"—the encoding for a set of traits and behaviors for playing the great game of survival. We might think of organisms *within* one species as "playing" one another, or organisms in *different* species as playing one another. As in Prisoner's Dilemma, the "payoff" to a given organism, with a given genotype, depends on the other organisms it encounters and plays. John defined the payoff of a strategy as its fitness. The average fitness of your genotype-strategy depends on who you encounter and play during your lifetime. The same holds for all organisms, hence genotypes, that are interacting.

Given this framework, you can guess what John did next. Each population of organisms may have the same genotype-strategy, or the population may have more than one. The shellfish population may have a variety of differently decorated shells; the population of starfish may have a variety of different sizes of beaks and suckers, tentacle shapes, and so forth. These genotype-strategies coevolve. At each generation, one or more genotypes in each of the coevolving populations undergoes mutations. Then these strategies compete, and the fittest ones spread most rapidly through the population. That is, the organisms "play one another," and the rate of reproduction of each genotype in the community of organisms is proportional to its fitness. Thus the population abundances of the fitter genotypes increase, while those of the less fit genotypes decrease. Both interacting populations within one species and populations between species coevolve this way.

John defined the concept of an evolutionary stable strategy as follows. At a Nash equilibrium, each player is better off not changing strategy as long as the other players play their own Nash equilibrium strategy. Similarly, an evolutionary stable strategy exists among a set of species when each has a genotype that it should selfishly keep as long as the other species keep their own ESS genotype. Each species has no incentive to change its strategy as long as the other species are "playing" their ESS strategies. If any species were to deviate, its own fitness would fall.

Evolutionary stable strategies are a fine idea. I'm glad I got to John's pneumonia, for he finally got rid of his negative rabbits and found the ESS idea. It is sometimes said that Morgenstern's major contribution to economics may have been to get von Neumann interested. It is not impossible that my own major contribution to evolutionary biology was a simple prescription for ampicillin.

Let's summarize something of the state of intellectual play at this point, for it represents the framework now in use by most population biologists and ecologists thinking about coevolution. Two main behaviors are imagined. The first image is of Red Queen behavior, where all organisms keep changing their genotypes in a persistent "arms race," and hence the coevolving population never settles down to an unchanging mixture of genotypes. The second main image is of coevolving populations within or between species that reach a stable ratio of genotypes, an evolutionary stable strategy, and then stop altering genotypes. Red Queen behavior is, as we will soon see, a kind of chaotic behavior. ESS behavior, when all species stop changing, is a kind of ordered regime.

Over the past decade or so, a great deal of effort has gone into attempts to understand whether and when Red Queen behavior or evolutionary stable strategies may occur in real coevolving ecosystems, and in attempts to understand what actually occurs out there in the wet, dry, flowered, fir-laden, tooth and clawed, snarling, sharing world of actual critters. The answers are still unknown, and indeed some coevolutionary processes may lie in the chaotic, Red Queen regime, while others may lie in the ordered evolutionary stable strategies regime. Over evolutionary periods of time, the very process of coevolution itself undoubtedly evolves, perhaps driving toward the Red Queen regime, perhaps toward the evolutionary stable strategies regime.

It is time to step up one level higher and consider the laws that may govern this evolution of coevolution. For perhaps there is a phase transition between the chaotic and the ordered regimes. Perhaps the evolution of coevolution tends to favor strategies that lie in this regime, near the edge of chaos.

The Evolution of Coevolution

To begin to think about the evolution of coevolution, we need a conceptual framework. I'll sketch it first, and then discuss it in more detail for the remainder of the chapter. As we have seen, coevolution concerns populations that are adapting on coupled fitness landscapes. The adaptive moves of one population clambering toward the peaks of its landscape deform the landscapes of its coevolutionary partners. As these deformations occur, the peaks themselves move. Adapting populations may succeed in climbing to the peaks and remaining on them, such that coevolutionary change ceases. Here is an ordered regime captured by Maynard Smith's evolutionary stable strategies. Alternatively, as each population climbs uphill, landscapes may deform so rapidly that all species chase after receding peaks forever in Red Queen chaos. Whether the coevolutionary process is in the ordered regime, the chaotic regime, or somewhere in between depends on the structure of the fitness landscapes and how readily each is deformed as populations move across them.

The *NK* models of fitness landscape discussed in Chapters 8 and 9 are a major help in thinking about why this happens. Think of two populations: frogs and flies. When some lucky frog, by quirk of mutational fortune, happens on a genetic mutation that gives it a sticky tongue, the sticky-tongue gene allele spreads like wildfire through the frog population. Thus the frog population will hop, so to speak, from the non-sticky-tongue allele to the sticky-tongue allele, hence hopping to a fitter genotype in its genotype space. Now if the fly population remained unchanged, the frog population would climb up its froggy fitness landscape to some local peak, and croak with joy.

Alas, the fly population, confronted with the newly arrived sticky-tongued frogs, finds that its own landscape is now deformed. Where once peaks of fitness existed, those peaks have shrunk, even perhaps to valleys. Flies should now respond to this novel frog gambit by developing slippery feet or even entire slippery bodies. New peaks in the fly genotype space emerge, many with "slippery feet" as a summit feature. Coevolution is a story of coupled dancing landscapes.

The fact that each landscape deforms when other partners move carries a powerful implication. On a fixed fitness landscape, at a low mutation rate a population will climb to a fitness peak and stop there. Mathematicians and physicists, describing flow on fixed fitness landscapes, say the system has a "potential function." For example, if we turn our fitness landscape upside down and rename our upside-down "fitness" as "energy," then physicists think of physical systems as lowering energy on simple or complex potential surfaces. When there is a fixed potential surface, the bottoms of local valleys, or local minima, or the tops of local fitness peaks, or maxima, are the natural end-point attractors of the system. A ball on a complex potential surface will come to rest at the bottom of some local valley.

But once the frog and fly populations start doing their stuff to each other, once the fitness landscape of each begins to deform as the other population moves on its own landscape, all bets are off. Neither population may ever come to rest on fitness peaks, for those peaks themselves may keep moving, and the adapting population may keep chasing peaks that forever elude them. Coevolving systems, thus, do not have a potential function. Mathematicians recognize such coevolving systems as general, complex, dynamical systems.

As a first rough statement, only two ultimate behaviors can occur in such a system. Each population is evolving across its own vast genotype space. If each population happens to climb to a fitness peak that is also consistent with the peaks located by all its coevolving partners, then all the populations will stop coevolving. By "consistent," I mean much the same idea as that captured by a Nash equilibrium, or evolutionary stable strategy. Each species at its own peak is better off not changing as long as the other species occupy their own peaks. The symbiosis between eukaryotic cells and mitochondrial endosymbionts is presumably an example of such mutual consistency, for the genetic couplings between cells and these intracellular organelles has presumably been stable for a billion years. This set of mutually consistent, locally optimal, unchanging genotypes is the analogue of a Nash equilibrium from game theory—like defect–defect in the Prisoner's Dilemma. Neither player has an incentive to change as long as the other player does not change.

The other possible behavior is that most or all species never settle down. They keep chasing receding peaks forever, forever doomed by their own best efforts to so deform the landscapes of their neighbors, and thus by indirect feedback loops, their own landscapes. All keep struggling uphill, Sisyphus-like, forever. This is Red Queen behavior.

So we see an ordered ESS regime, where species are frozen and unchanging on local peaks, and a chaotic Red Queen regime, where species ceaselessly surge across their genotype spaces. With Boolean models of genetic nets we also saw an ordered regime, with its frozen components spanning the network, and a chaotic regime. In the case of the genetic networks, we found a *continuum,* an axis between order and chaos. And we found evidence that the most complex computation might occur in the ordered regime near the phase transition between order and chaos, the edge-of-chaos "complex regime." In the ordered regime near the phase transition, complex but nonchaotic cascades of activities can propagate across a network, allowing coordination of complex sequences of events. Might we find a similar continuum between order and chaos in coevolving systems?

I am now about to tell you that such a continuum exists, that between the ordered ESS regime and chaotic Red Queen regime lies something like the phase transition we've called the edge of chaos. It begins to appear that the evolution of coevolution "prefers" this phase transition. An ecosystem deep in the ordered ESS regime is too rigid, too frozen in place, to coevolve away from poor local peaks. In the Red Queen chaotic regime, on the contrary, species climb and plunge on heaving fitness landscapes and therefore have low overall fitness. Fitness, it turns out, can be highest at an intermediate position on the order–chaos axis, near the poised phase transition. How do ecosystems get to the poised edge-of-chaos state? Evolution takes them there.

Whether a coevolving system lies in the ordered or the chaotic regime depends on the ruggedness of the landscape explored by each species, and how much each species's landscape is deformed by the adaptive moves of its partners. If each landscape has few peaks and the locations of those peaks are drastically altered when other species move on their own landscapes, then species rarely "catch up" to the receding peaks on their own landscapes. The system is in the Red Queen chaotic regime. In contrast, if each landscape has many peaks and those peaks do not move very much when other species move on their own landscape, then each species can climb to peaks quite easily. The system is in the ordered ESS regime. Pretty obviously, then, whether the coevolutionary system is in the ordered or the chaotic regime depends on landscape structure and deformability. But this then leads to higher-level questions.

What governs landscape structure and deformability? Is there a "good" structure and deformability such that the coevolving partners can all coevolve as well as possible? If so, we might find that there are laws governing the evolution of coevolution—laws that all but guarantee ecosystems that ensure the best possible living for all.

Coupled Fitness Landscapes

Recall that in the *NK* model of rugged fitness landscapes, *N* stands for the number of traits in an organism, or genes in its genome. Any trait, or gene, can come in several versions, or alleles. In a simple case, each trait or gene comes in only two alleles, 1 and 0: 1 might stand for blue eyes; 0, for brown eyes. We modeled "epistatic interactions" among genes, by saying that any gene's fitness contribution depended on the 1 or 0 state of that gene and the 1 or 0 states of *K* other genes. Here *K* measures the epistatic coupling among the genes within one organism, how interdependent they are. Finally, we modeled the "fitness contribution" a given gene made to the entire genome, or organism: for each of the combinations of states of that gene and its *K* epistatic inputs, we assigned a random decimal number between 0.0 and 1.0. Finally, we defined the fitness of the whole genotype as the average of the fitness contributions of each gene. The result was a fitness landscape.

Now how shall we begin to think about coevolution? When the frog develops its sticky tongue, that tongue impinges on the fitness of the fly because of specific traits of the fly. We have considered the feet of the fly, nonslippery or slippery, as one such trait. But other traits may matter in the response to the sticky tongue: bad-tasting flies, flies with fast-acting sticky-stuff dissolver, flies whose sense of smell is specially tuned to detect the chemicals in the sticky stuff so the flies can fly away, and so forth. That is, it is natural to think that a specific trait of the frog impacts on the fitness of the fly through several of the N traits of the fly. Conversely, some of the fly traits affect the fitness of the frog through several specific frog traits. The fleeter-flying fly decreases the fitness of the frog. Perhaps the frog counters with a longer tongue, a faster-flicking tongue, a less strongly smelling sticky stuff, and so forth. Just as we can use the NK model to show how genes interact with genes or how traits interact with traits within one organism, we can also use it to show how traits interact with traits between organisms in an ecosystem.

Let's build a simple model of organisms whose fitness landscapes are coupled to one another. Suppose that each of the N traits in the fly, as before, makes a fitness contribution that depends on K other traits in that fly, but also depends on the states, 1 or 0, of C traits in the frog. Likewise, each of the N traits in the frog makes a fitness contribution that depends on K other traits in the frog, but also on the states, 1 or 0, of C traits in the fly. Next we need to couple the traits of the frog and the fly. The easiest way to do this is to say that the fitness contribution of each trait in the fly depends on that trait, 1 or 0; on the K other traits in the fly, 1 or 0; and *also* on the C traits in the frog, 1 or 0. Then, for all possible combinations of all these inputs, assign a random decimal number between 0.0 and 1.0 describing the contribution the fly trait in question makes to the overall fitness of the fly.

When we carry this out for all the N traits of the fly and the N traits of the frog, the fly and frog fitness landscapes are coupled. A move by the fly up its landscape alters the 1 and 0 pattern of its N traits, and hence deforms the fitness landscape of the frog. Reciprocally, let the frog population move on the frog landscape, and the fly landscape will deform.

We need two more simple steps. How many species should we have in our model ecosystem? And how many species should each species be coupled to and in what pattern? Well, to come to grips with this in a starting model, it suffices to begin with a pretty odd-looking ecosystem. Let there be 25 species, arranged on a 5 × 5 square lattice, like square tiles, such that each species is connected to its four neighbors to the north, south, east, and west.

Then, to actually make a computer simulation of such a model, we need a few more details. Here's one way of proceeding. We will treat each species as though its entire population were genetically identical. At each "generation," the population will seek a fitter genotype by mutating a single, randomly chosen gene to the alternative allele. If the new mutant genotype is fitter, the population will move to this new point on its landscape. So the population will perform an adaptive walk, remaining unchanged or taking one step "uphill" per generation. Each species will have one chance to make an adaptive step at each generation. So "generations" measures time elapsed. Here is what happens.

The ecosystem tends to settle into the ordered, evolutionary stable strategies regime if either epistatic connections, K, within each species are high, so that there are lots of peaks to become trapped on, or if couplings between species, C, is low, so landscapes do not deform much at the adaptive moves of the partners. Or an ESS regime might result when a third parameter, S, the number of species each species interacts with, is low, so that moves by one do not deform the landscapes of many others (Figures 10.3 and 10.4).

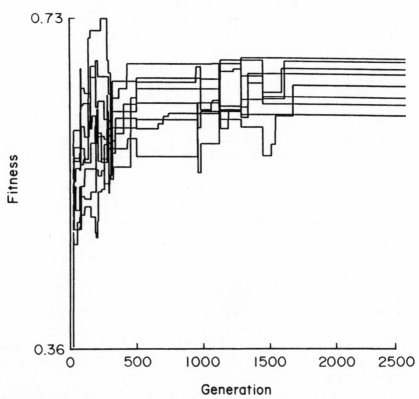

N=24 K=13 Adjacent C=1 Populations=8

Figure 10.3 *An evolutionary stable strategy (ESS). Coevolution among eight species, each governed by an NK landscape. Each of the N traits in each species is affected by C = 1 traits in each of the seven other species. The system reaches a steady state by about generation 1,600, the sign of an evolutionary stable strategy.*

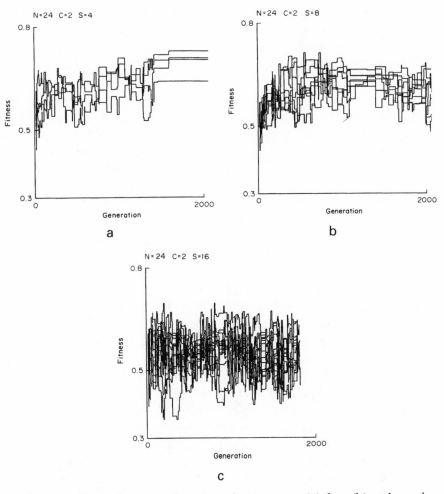

Figure 10.4 *The Red Queen effect. Coevolution among* (a) *four,* (b) *eight, and* (c) *16 species. Note that as the number of species increases, the mean fitness decreases and the fluctuations in fitness increase. For* (b) *and* (c) *no evolutionary stable equilibrium was found in 8,000 generations. These systems remained in chaotic Red Queen behavior.*

There is also a chaotic Red Queen regime where the species virtually never stop coevolving (Figure 10.4c). This Red Queen regime tends to occur when landscapes have few peaks to get trapped on, thus when K is low; when each landscape is deformed a great deal by adaptive moves of other species, thus when C is high; or when S is high so that each species is directly affected by very many other species. Basically, in this case, each species is chasing peaks that move away faster than the species can chase them.

At first, it might seem surprising that low K leads to chaotic ecosystems; in the NK Boolean networks, high K led to chaos. The more intercouplings, the more likely a small change was to propagate throughout and cause the Boolean system to veer off into butterfly behavior. But with coupled landscapes it is the interconnectedness between the species that counts. When intercoupling, C, is high, moves by one species strongly deform the fitness landscapes of its partners. If any trait in the frog is affected by many traits in the fly, and vice versa, then a small change in traits of one species alters the landscape of the other a lot. The system will tend to be chaotic. Conversely, the ecosystem will tend to be in the ordered regime when the couplings between species, C, is sufficiently low. For much the same reason, if we were to keep K and C the same, but change the number of species S any one species directly interacts with, we would find that if the number is low the system will tend to be ordered, while if the number is high the ecosystem will tend to be chaotic (Figure 10.4).

Your antenna might twitch a bit now. If there is a chaotic regime for one set of parameters and an ordered regime for another, then what happens as we tune the parameters? How does one get from order to chaos? And for what parameter settings of the coevolving ecosystem is the average fitness of all the species highest?

The first interesting result is that as the parameters, K and C, and number of partners, S, are changed along the axis, from the chaotic regime to the ordered regime, the average fitness at first *increases* and then *decreases*. The highest average fitness of the players occurs when the position on the order–chaos axis is set to a *middle position,* neither deeply chaotic nor deeply ordered. Figure 10.5 shows this for simulations in which only the richness of epistatic interactions within species, K, changes, while C and the number of partners, S, are held constant. Why should positions at the extremes of the order–chaos axis lead to low fitness? Deep in the chaotic regime, fitnesses rise and plunge so chaotically that average fitness is low (Figure 10.4c). Deep in the ordered regime, K is very high: with its genome so densely interconnected, each organism is caught in a web of conflicting constraints that results in many low-fitness peaks representing very poor compromises. The fitness peak that each species clings to is a very poor peak. Thus deep in the ordered regime, average fitness is low. So fitness is maximized for an intermediate position on the order–chaos axis.

In fact, the results of our simulations suggest that the very highest fitness occurs *precisely* between ordered and chaotic behavior! How can we tell? We can see how deeply ecosystems are in the ordered regime by seeing how readily they "freeze" into evolutionary stable strategies. Just start 100 similar ecosystems, and watch when each of these reaches an

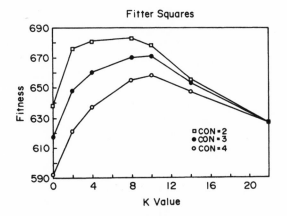

Figure 10.5 *Tuning an ecosystem. As the richness of epistatic connections between species, K, is increased, tuning the ecosystem from the chaotic to the orderly regime, average fitness at first increases and then decreases. It reaches the highest value midway between the extremes. The experiment is based on a 5 × 5 square lattice ecosystem, in which each of 25 species interacts with at most four other species. (Species on the corners of the lattice interact with two neighbors [CON = 2]; species on the edges interact with three neighbors [CON = 3]; and interior species interact with four neighbors [CON = 4]. N = 24, C = 1, S = 25.)*

ESS and stops coevolving. After some number of generations, 50 similar ecosystems have frozen into their own ESS (Figure 10.6). Take this number of generations as an estimate of the average freezing time for such model ecosystems. In the chaotic regime, this half-time is very long indeed; in the ordered regime, it is very short. Figure 10.6 shows our model ecosystem "run" for 200 generations. The couplings between species, C, is held constant at 1. Each species is coupled to its north, south, east, and west partners on our 5 × 5 lattice ecosystem. As epistatic interactions within each species, K, is tuned from low, $K = 0$, to high, $K = 22$, ecosystems are tuned from the chaotic to the ordered regime. Figure 10.6 shows that when K is high, model ecosystems freeze onto ESS equilibria rapidly. When K is 8 or less, the ecosystem is chaotic and none of the 100 similar ecosystems attain ESS during the 200 generations. The critical thing to notice is this: when K is increased from 8 to 10, a modest fraction of the ecosystems freeze into their ESS over the 200 generations. So over a 200-generation time scale, ecosystems are just starting to freeze, they are just between ordered and chaotic behavior, when our parameter of epistatic interactions, K, is set to $K = 10$.

Now if we compare Figures 10.5 and 10.6, we can see that the middle

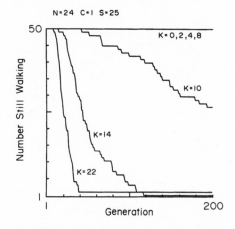

Figure 10.6 *The edge of chaos. The fraction of 5 × 5 ecosystems that have not yet become frozen into evolutionary stable equilibria is plotted against the elapsed number of generations. Note in the chaotic regime (K < 10) none of the ecosystems freeze into an ESS in the 200 generations available. In the orderly regime (K ≥ 10), some or most of the systems freeze at Nash equilibria, and do so more rapidly as K increases. Thus for K = 10, ecosystems are at the edge of chaos, just beginning to freeze in the 200 generations available.*

position on the chaos–order axis where average fitness is maximized is just at $K = 10$, just precisely between chaotic and ordered behavior! The highest average fitness occurs precisely at the transition from order to chaos. Deep in the ordered regime, fitness peaks are low because of conflicting constraints. Deep in the chaotic regime, fitness peaks are high, but are too few and move too rapidly to be climbed. The transition regime occurs precisely at that position on the axis where the peaks can just be climbed on the time scale available. Here the peaks are simultaneously the highest possible and still attainable in the time available. So the transition between order and chaos appears to be the regime that optimizes average fitness for the whole ecosystem.

Are there hints here of a general law? Is the transition zone between order and chaos the "good" regime for a coevolutionary system to be in? And can an evolutionary process, natural selection acting on individuals, an invisible hand, automatically tune a coevolving system to this regime? We are ready for the final step—the evolution of coevolution.

My young colleague Kai Neumann and I have been looking into the conditions under which coevolving species will spontaneously evolve to the regime of highest average fitness. Let's imagine our model ecosystem as one in which N and S are fixed; the number of species and the number of species each interacts with remains constant. We want, as usual, to allow each species to evolve on its own deforming fitness land-scape. In addition, we want each species to be able to *evolve the rugged-ness of its landscape* by altering its own internal epistatic coupling level, K. Finally, we want to allow extinction to occur. To do this, we imagine that at each generation, each species in the ecosystem can experience one of four things:

1. It can stay the same.

2. It can mutate a single gene, moving to a neighboring point on the landscape.

3. It can mutate by increasing or decreasing K for each of its genes by 1, altering the ruggedness of its landscape.

4. Some other randomly chosen species in the ecosystem can send a copy of itself over to try to "invade" the niche of the species in question.

All four possibilities are tried out with the north, south, east, and west coevolutionary partners. Whichever of the four possibilities results in the highest fitness wins:

1. If the unmutated species in its "niche," as defined by its par-ticular north, south, east, and west neighbors, is fittest, no change occurs.

2. If the familiar single-mutant neighbor is fittest, the species moves on its landscape.

3. If the mutation altering K is fittest, the species keeps the same genotype, but alters the ruggedness of its landscape.

4. If the invader is fittest, the initial species goes extinct and the invader occupies its niche, thereafter coevolving with the north, south, east, and west species.

Its a pretty cutthroat world in silico.

One might start the system with all species having very high K values, coevolving on very rugged landscapes, or all might have very low K val-ues, coevolving on smooth landscapes. If K were not allowed to change, then deep within the high-K ordered regime, species would settle to ESS rapidly; that is, species would climb to poor local peaks and cling to them. In the second, low K, Red Queen chaotic regime, species would never attain fitness peaks. The story no longer stops there, however, for the species can now evolve the ruggedness of their landscapes, and the persistent attempts by species to invade new niches, when successful, will insert a new species into an old niche and may disrupt any ESS attained.

What happens when we run the model is somewhere between intriguing and amazing. However you start the model ecosystem, with any distribution of K values across the species, the system appears to converge to an optimal intermediate K value, where average fitness is highest and average rate of extinction is lowest! The coevolving system tunes its own parameters, as if by an invisible hand, to an optimal K value for everyone.

Figures 10.7 and 10.8 show these results. Each species has $N = 44$ traits; hence epistatic coupling can be as high as 43, creating random landscapes, or as low as 0, creating Fujiyama landscapes. As generations pass, the average value of K in the coevolving system converges onto an intermediate value of K, 15 to 25, and stays largely within this narrow range of intermediate landscape ruggedness (Figure 10.7). Here fitness is high, and the species do reach ESS equilibria where all genotypes stop changing for considerable periods of time, before some invader or

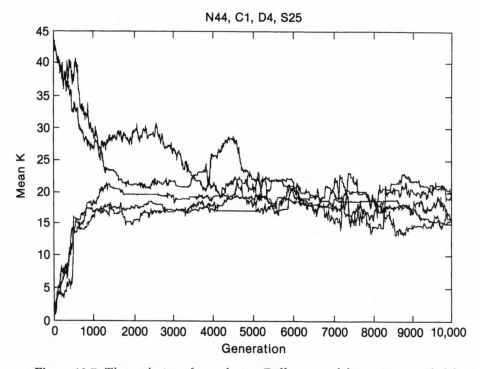

Figure 10.7 *The evolution of coevolution. Different model ecosystems with different initial average values of* K *are allowed to evolve. As generations pass, the value of* K *converges onto the intermediate range, between order and chaos. As a result, mean fitness increases, and the frequency of extinction events decreases. As if by an invisible hand, the system tunes itself to the optimal* K *values for everyone.*

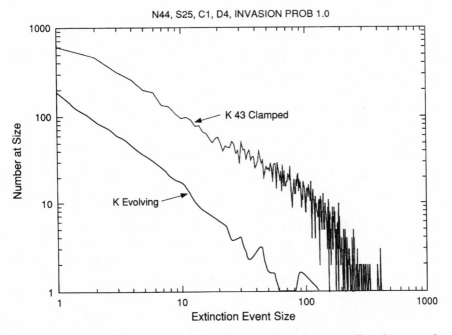

Figure 10.8 *Controlling extinction events. Using logarithmic scales, the sizes of hypothetical extinction events are plotted against the number of events of each size. The result is a power-law distribution. In the upper distribution, an ecosystem is held artificially deep in the high-K regime, K = 43. Fitness is low in this regime because of conflicting constraints, so huge avalanches of extinction events cascade through the system. The lower power-law distribution corresponds to a system where landscape ruggedness has been allowed to evolve to an intermediate K range, K = 22. Fewer and smaller extinction avalanches occur.*

invaders disrupt the balance by driving one or more of the coadapted species extinct.

When one species is driven extinct, the event may trigger a small or a large extinction avalanche that sweeps through some or all of the ecosystem. Why? When one species goes extinct, it is replaced by an invader. The invader is new to the niche, is typically not at a local peak, and therefore adapts in new ways that change its genotype. These moves change the fitness landscapes of its north, south, east, and west neighbors, typically lowering their fitness. As their fitness is lowered, they become more readily subject to successful invasion and extinction. So the avalanches of extinction tend to spread outward from any single extinction event.

The extinction avalanches, measured in the number of species that go extinct in an avalanche, appears to be a power-law distribution (Figure

10.8). If we plot the logarithm of the size of an extinction avalanche on the x-axis and the logarithm of the number of avalanches at each avalanche size on the y-axis, we get a more or less straight line, with many small avalanches and few large ones. Indeed, even if K is held at a fixed high or low value, the distribution of extinction events for such model ecosystems appears to be a power law.

When K is held high or low, deep in the ordered regime or deep in the chaotic regime, huge extinction avalanches rumble through the .model ecosystems. The vast sizes of these events reflect the fact that fitness is low deep in the ordered regime because of the high-K conflicting constraints, and fitness is low deep in the chaotic regime because of the chaotic rising and plunging fitness values. In either case, low fitness of a species makes it very vulnerable to invasion and extinction. The very interesting result is that when the coevolving system can adjust its K range, it self-tunes to values where average fitness is as high as possible; therefore, the species are least vulnerable to invasion and extinction, so extinction avalanches appear to be as rare as possible. This shows up in Figure 10.8, which compares the size distribution and total number of extinction events deep in the ordered regime and after the system has self-tuned to optimize landscape ruggedness, K, and fitness. After the ecosystem self-tunes, the avalanches of extinction events remain a power law—the slope is about the same as when deep in the ordered regime. But over the same total number of generations, far fewer extinction events of each size occur. The self-tuned ecosystem also has far fewer extinction events than does an ecosystem deep in the chaotic regime. In short, the ecosystem self-tunes to minimize the rate of extinction! As if by an invisible hand, all the coevolving species appear to alter the rugged structures of the landscapes over which they evolve such that, on average, all have the highest fitness and survive as long as possible.

We are not sure how this invisible hand works. But here is our current best guess: we have seen that over a 200-generation time scale, average fitness was optimized at a value of K that allowed species to just reach the highest peaks in the time available. At that time scale, $K = 10$ ecosystems climbed the highest peaks and managed to reach ESS, after which the species froze at mutually consistent local peaks and stopped changing. In that case, it was Kai and I who set the 200-generation time scale. In the present studies, we are setting no time scale. The model ecosystem with extinctions is setting its own time scales by itself.

How is this done? When species invade niches and drive old species extinct, the new species tend to disrupt any evolutionary stable strategy the old species may have located. Thus the system that would be frozen at an ESS were no extinction events to occur "unfreezes" itself. We suspect that our model sets its own time scales in an inevitable and self-consistent way. It self-tunes to the landscape ruggedness where species can just find the highest peaks and attain ESS *slightly faster* than the average rate at which the ecosystem is disrupted by extinction events. If landscape ruggedness were greater because K were higher, species would reach ESS equilibria more rapidly, but be more vulnerable to extinction because peaks would be lower. Were landscapes smoother and peaks higher because K were lower, species would be oscillating chaotically and not reach ESS in the time available, and so would be more vulnerable to extinction. Just at the poised position where, on average, species are climbing to ESS slightly faster than the ESS organization is being disrupted by episodic extinction events, fitness is optimized and the extinction rate is minimized.

An invisible hand is in sight. Candidate laws governing the evolution of coevolution are not impossible. Over evolutionary time, ecosystems may self-tune to a transition regime between order and chaos, maximizing fitness, minimizing the average extinction rate, yielding small and large avalanches of extinctions that ripple or crash through the ecosystem. We are all but players who strut and fret our hour upon the stage, and then are heard no more. But we may collectively, and blindly, tune the stage so each will have the best chance at a long, long hour.

Sandpiles and Self-Organized Criticality

In 1988, a simple, beautiful, perhaps even correct theory was born. I heard enthusiastic rumors of sandpiles and self-organized criticality, of physicists Per Bak, Chao Tang, and Kurt Wiesenfeld, from Phil Anderson and the other solid-state physicists at the Santa Fe Institute. Finally Per Bak came to visit from Brookhaven National Laboratories. We began what has become a warm friendship with an argument. Bak kept telling me that autocatalytic sets could not work. Two lunches and a bit of yelling later, I had convinced him that they could. Per Bak is not easy to convince, but worth the effort, for he is very creative. He and his friends may have discovered a very general phenomenon, one that ties to many of the themes of this book: self-organized criticality.

We are to picture a table. Above, rather like the hand of God outstretched toward Adam's on the ceiling of the Sistine Chapel, a hand is poised, holding sand. Sand slips persistently from the hand onto the table, piling higher and higher until the heap of sand slides in avalanches from the top of the table onto a distant floor.

Piling higher, then reaching its rest angle, the sandpile achieves a rough stationary state. As the sand is trickled onto the pile, many small sandslides avalanche down the sides, and a few large sandslides, massive and grave, but for the childlike character of a mere table heaped with sand, make their way to the floor.

As we have noted before, if one plots these avalanches, one finds our now familiar power-law distribution. The distribution of sizes of sand-slide avalanches looks like the distribution of sizes of extinction events in our model ecosystem (Figure 10.8). There are many small avalanches and few large ones. Sandslide avalanches can arise *on all scales*. For many things—human height, the loudness of a frog's croak—there is a typical size. But there is no typical size for a sandpile avalanche. Estimate the mean height of humans, and the estimate gets *better* the more individuals we measure. Estimate the mean size of avalanches, and the estimate gets *bigger* the more examples we measure. The largest scale is set by only the size of the table. Were the table enormous and piled high with an enormous sandpile, one would get many little avalanches and fewer big avalanches, but if one waited long enough, one would get rare massive avalanches crashing down. Bigger and bigger avalanches, on all possible scales, occur, but at rarer and rarer intervals, as specified by the power law. Furthermore, the size of the avalanche is unrelated to the grain of sand that triggers it. The same tiny grain of sand may unleash a tiny avalanche or the largest avalanche of the century. Big and little events can be triggered by the same kind of tiny cause. Poised systems need no massive mover to move massively.

This is what Bak, Tang, and Wiesenfeld mean by self-organized criticality. The system tunes itself, as if by an invisible hand, to the critical-rest angle of sand and remains poised there as sand is forever trickled onto the pile from above.

Bak and his colleagues argue that many features of the physical, biological, and perhaps even economic worlds may exhibit self-organized criticality. For example, the size distribution of earthquakes, measured on the familiar Richter scale, which is just the logarithm of the total energy released in the quake, follows a power law, with many small quakes and few large ones. The floodings of the Nile follow a similar power law, with many small floods and rare large ones. (I discuss potential economic examples later.) Bak even wonders whether the clustered nature of matter in the universe, galaxies and clusters of galaxies, shows a similar power law, with many small clusters and few large ones, because the universe is poised between expanding forever and eventually falling back on itself into the Big Crunch. We have seen other evidence of poised systems and power laws in this book: local ecosystems may be poised between subcritical and supracritical behavior, with power-law avalanches of molecular novelty. Genomic networks may evolve to the ordered regime near the poised edge of chaos, with power-law avalanches of changing gene activities. And we have just seen evidence that, as if by an invisible hand, ecosystems may self-tune their structure to a poised regime between order and chaos, with power-law distributions of extinction events.

Is this view right? We do not know yet. But there are clues, perhaps very important ones, that something like the picture I have just painted may be correct.

Some of the clues concern the size distribution of extinction events. These do, roughly, appear something like power laws—but not quite. Data are known in at least two different domains: real extinction events witnessed in the rocks, and model extinction events in an artificial-life simulation by Tom Ray (ecologist, student of Costa Rican forests, and Santa Fe Institute regular).

David Raup has analyzed the entire 550 million years from the Cambrian, called the Phanerozoic, into 77 periods of about 7 million years each. For each period, Raup assembled the data for the number of families that went extinct. He found many small extinction events and few large ones. Figure 10.9 shows the size distribution and the log–log plot of the size distribution. As you can see, one does not get a straight line;

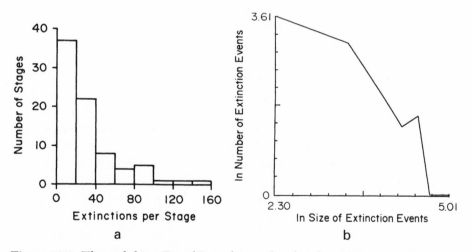

Figure 10.9 *The real thing. David Raup has analyzed real extinction events according to the number of families that went extinct. (a) His data show many small extinction events and few large ones. (b) When the data are replotted in natural logarithmic form, the result is not quite a linear power-law distribution; there are fewer larger extinctions than might be expected.*

the curve bends over (Figure 10.9*b*). One does not quite get a power-law distribution. There are fewer large extinction events than one might expect compared with the number of smaller extinction events.

What causes the large and small extinction events? Well, a humdinger occurred at the end of the Cretaceous and, many think, wiped out the dinosaurs. It appears that a massive meteor waded into the Atlantic just off the Mexican coast where the Maya later flourished. The evidence for this meteor is very good. Some scientists think that all extinction events are caused by such exogenous catastrophes. Little extinction events would be caused by little meteors; big extinction events would be caused by large meteors. They may be right. Conversely, most extinction events may reflect endogenous processes within ecosystems, such as those we have modeled. Indeed, both small and large extinction events may be triggered by the same small initial events. Our results suffice to indicate that we are not forced by the existence of small and large extinction events to posit small and large causes of those events.

If most extinction events mirror internal processes in the biosphere, then Raup's data are consistent with ecosystems rather more in the ordered regime than at a phase transition between order and chaos. Conversely, 77 data points is not very much on which to base a conclusion. Data assembled at the level of genera or species, rather than families, would be very helpful. So. A clue, not a proof, is now witnessed in the record.

Tom Ray created an artificial world, Tierra, to study ecology in silico. He developed computer programs that "live" in the core of the computer. Each program is able to copy itself onto an adjacent location in the core. And programs can mutate, forming a community with one another. Lively things happen on this silicon stage. Critters emerge that, like any good parasite, borrow the computer instructions of their hosts to replicate themselves. Hosts respond by forming boundaries to ward off the parasites. The parasites evolve into hyperparasites, which sneak past the defensive walls of the hosts.

But in time, Tom's critters croak. They disappear from his stage, and are heard no more. I asked Tom to plot the size distribution of extinction events in Tierra. Figure 10.10 shows the log–log plot of the results: it is nearly a power law, with many small and few large extinction events. As with Raup's data, the curve bends over. There are rather too few massive extinction events. This may reflect "finite size" effects, for Ray's core is only so big, so avalanches can be only so large. Come to think of it, the earth is only so big, and extinction avalanches can be only so large. The Permian extinction wiped out 96 percent of all species, it is said. One cannot get much bigger than that. Perhaps the

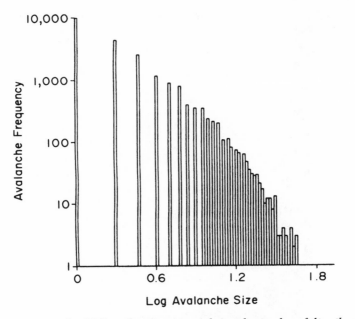

Figure 10.10 *Artificial life and extinction. A logarithmic plot of distribution of sizes of extinction events in Tom Ray's Tierra simulations.*

bowed-curve deviation in Raup's data reflects the finite size of the entire biosphere (Figure 10.9*b*).

Another feature that we can study also provides a clue. Because of the small and large extinction events and the speciation events that precede them, each species, and hence each genus or higher taxon, has a definite life span. The species began at some point and went extinct at some point. The genus is initiated with its first species, and goes extinct when its last species is extinguished. Raup has analyzed data on vertebrate and invertebrate marine animals gathered by paleontologist Jack Sepkowski representing 17,500 genera and about 500,000 species. Raup used this to show the distribution of life spans of genera. Figure 10.11 shows Raup's data. The distribution falls off rapidly; most genera die "young," but a long "tail" extends to the right. Some genera last 150 million years or more. Figure 10.12 shows the corresponding distribution of life spans of the model species from the simulations Kai Neumann and I performed. The distributions are strongly similar; both fall off rapidly, but extend a long tail. Most model species die young; a few survive a very long time. We do not yet know that ecosystems over evolutionary time coevolve to a self-tuned edge-of-chaos regime, but the parallels between the size of extinction events and the life-span distributions are supportive clues.

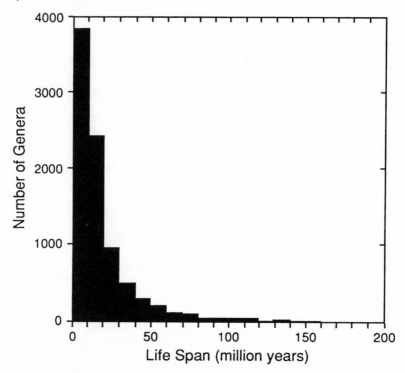

Figure 10.11 *Infant mortality in marine life. The number of fossil vertebrate and invertebrate marine general arranged according to their life spans. Most genera die young, but a long tail extends to the right.*

What applies to organisms may also apply to artifacts. If both evolve on rugged landscapes because of conflicting constraints, as I suggested in Chapter 9, both also *coevolve*. Firms and technologies also strut and fret their hour upon the stage, and then are heard no more. The arrival of the automobile ensured the extinction of the horse as a means of transport. With the horse went the buggy, the buggy whip, the smithy, the saddlery, the harness maker. With the car came the oil and gasoline industry, motels, paved roads, traffic courts, suburbs, shopping malls, and fast-food restaurants. Organisms speciate and then live in the niches created by other organisms. When one goes extinct, it alters the niche it helped create and may drive its neighbors extinct. Goods and services in an economy live in the niches afforded by other goods and services. Or rather, we make a living by creating and selling goods and services that make economic sense in the niches afforded by other goods and services. An economy, like an ecosystem, is a web of coevolving agents. Austrian economist Joseph Schumpeter spoke of gales of creative destruction. During such gales, new technologies come into existence and old ones are driven to extinction.

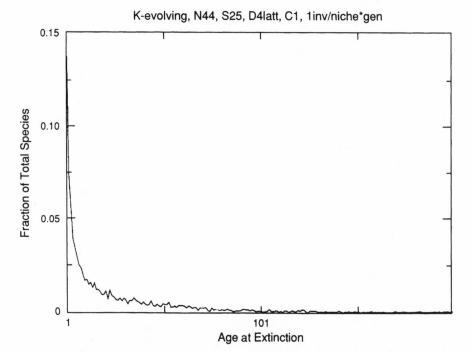

Figure 10.12 *Infant mortality in artificial life. The age that simulated species in a model by Kauffman and Neumann went extinct, plotted against the fraction of total species becoming extinct at each age. The distribution is similar to that of Figure 10.11: most die young; a few live a very long time.*

One knows that small and large avalanches of technological change propagate through economic systems. Do these avalanches follow a power-law distribution, with many small and few large avalanches? It seems reasonable, but I do not know of detailed work on the size distribution of such avalanches. On a smaller scale, economists talk about the death rate of firms as a function of years since founding. It is well known that "infant mortality" is high. The older the firm, the more likely it is to survive until the next year. No accepted theory yet accounts for such mortality curves. But we might notice that the coevolutionary model Neumann and I are studying fits these facts. Figure 10.13 shows the probability of mortality as a function of age in our model ecosystem. Infant mortality is high and falls as species age. Why? Think of an invader, when invading a niche, as the founding of a new species or firm. At first, the new firm is of low fitness in its new niche, and hence is likely to be displaced in its turn by another new invader. Moreover, the coevolutionary turmoil it induces in its vicinity may keep its fitness low for a while, leaving it prey to extinction in its youth. But as it and its region of the ecosystem climb toward ESS equilibria, the

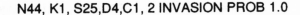

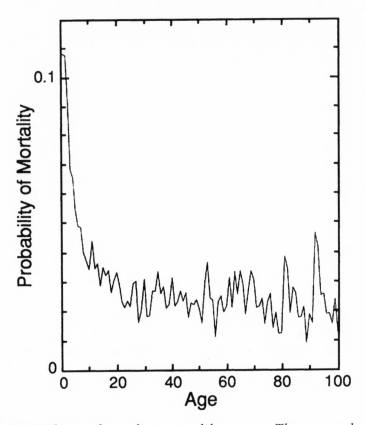

Figure 10.13 *The age of mortality in a model ecosystem. The younger the "species," the higher the chance of extinction.*

firm's average fitness increases. It becomes less subject to successful invasion and extinction. Thus its death rate per generation decreases until an ESS is reached. Ultimately, someone will probably succeed in invading, and the firm will go the way of all flesh. For this schematic explanation to fit economic facts, of course, fitness needs a direct economic interpretation such as capitalization or market share or other measures of success. As firms mature, they tend to become well capitalized and capture market share. Once established, firms are not rapidly driven out of business.

Organisms and artifacts may evolve and coevolve in very similar ways. Both forms of evolution, that crafted by the blind watchmaker and that crafted by us, the mere watchmakers, may be governed by the same global laws. Perhaps small and large avalanches of extinction events, as trumpeted by Schumpeter, rumble through economic systems with a power-law distribution. Perhaps small and large avalanches of extinction events rumble through the biosphere. Perhaps we all make our lives together, blindly self-tuning the games we mutually play, the roles we mutually fulfill, to a self-organized critical state. If so, Per Bak and his friends will have found a very wonderful law, for as we've suggested, a new philosophy of life tiptoes on their mathematical sandpile. Our smallest moves may trigger small or vast changes in the world we make and remake together. Trilobites have come and gone; *Tyrannosaurus* has come and gone. Each tried; each strode uphill; each did its evolutionary best. Consider that 99.9 percent of all species have come and gone. Be careful. Your own best footstep may unleash the very cascade that carries you away, and neither you nor anyone else can predict which grain will unleash the tiny or the cataclysmic alteration. Be careful, but keep on walking; you have no choice. Be as wise as you can, yet have the wisdom to admit your global ignorance. We all do the best we can, only to bring forth the conditions of our ultimate extinction, making way for new forms of life and ways to be.

Well, here is a pretty philosophy, Horatio. Do your best, get up in the morning, grab some coffee, hit the cornflakes, rush through traffic to the subway, crowd into the elevator up to the office, pile through the piles of everyone else's already sent piles, climb whatever ladder your career choice sent you up, and in the end, what do you get? Brief hour upon the stage. Tough, pal, but the Ediacrin flora and fauna had it no better.

Here is no Panglossian world, or Hobbsian either. Perhaps here is the reality we have always suspected. Do your best; you will ultimately slip into history along with the trilobites and other proud personae in this unfolding pageant. If we must eventually fail, what an adventure to be players at all.

Chapter 11

In Search of
Excellence

S idney Winter, an economist now at the Wharton School at the University of Pennsylvania after a four-year stint as chief economist at the General Accounting Office, spoke at a recent Santa Fe Institute meeting on "Organizational Evolution" and immediately captured our attention. After all, most of us were academic scientists. Sid spoke from experience near the center of the United States government about global changes in our economic life. "There are four horsemen of the workplace," he said.

Technology, Global Competition, Restructuring, Defense Conversion. These are dominating the post–Cold War period. We need jobs, good jobs, but we don't know how to be certain that the economy will generate those jobs. Health reform, welfare reform, and trade policy are around the corner. We neither know how to achieve these nor understand their impact. The de facto tenure of jobs in U.S. firms is decreasing. Companies are out-sourcing. Rather than performing all parts of the total job inside the firm, many subtasks are being purchased from other firms, often in other countries. This is leading to disintegration of the vertical organization of firms. Mergers and acquisitions are pulling old companies into new forms, then spinning out components into new structures. Trade liberalization is upon us. We are downsizing. It is all captured by a common theme: repackaging. We are shifting packages of economic activity into new smaller units. The folk model of organization as top–down and centralized is out of date. Organizations are becoming flatter, more decentralized.

I listened with surprise. Organizations around the globe were becoming less hierarchical, flatter, more decentralized, and were doing so in the hopes of increased flexibility and overall competitive advantage.

Was there much coherent theory about how to decentralize, I wondered. For I was just in the process of finding surprising new phenomena, a new edge-of-chaos story, that hinted at the possibility of a deeper understanding of how and why flatter, more decentralized organizations—business, political, and otherwise—might actually be more flexible and carry an overall competitive advantage.

A few weeks later, just as I was absorbing this message, the Santa Fe Institute held an "outpost" meeting at the University of Michigan. The aim was to connect the work on the "sciences of complexity" going on in Santa Fe with that of colleagues at the University of Michigan. Computer scientist John Holland, who has had a major impact with his development of "the genetic algorithm," which uses landscape ideas, mutation, recombination, and selection to solve hard mathematical problems, is the glue between the institute and his home institution in Ann Arbor. The dean of the Department of Engineering, Peter Banks, was charismatic. "Total Quality Management is taking over, integrating new modular teams in our firms," he said. "But we have no real theoretical base to understand how to do this well. Perhaps the kind of work going on in Santa Fe can help." I nodded my head vigorously, hopeful, but not necessarily convinced.

Why would I, the other scientists at Santa Fe, or our colleagues around the globe studying complexity be interested in potential connections to the practical problems of business, management, government, and organizations? What are biologists and physicists doing poking into this new arena? The themes of self-organization and selection, of the blind watchmaker and the invisible hand all collaborating in the historical unfolding of life from its molecular inception to cells to organisms to ecosystems and finally to the emergent social structures we humans have evolved—all these might be the locus of law embedded in history. No molecule in the bacterium *E. coli* "knows" the world *E. coli* lives in, yet *E. coli* makes its way. No single person at IBM, now downsizing and becoming a flatter organization, knows the world of IBM, yet collectively, IBM acts. Organisms, artifacts, and organizations are all evolved structures. Even when human agents plan and construct with intention, there is more of the blind watchmaker at work than we usually recognize. What are the laws governing the emergence and coevolution of such structures?

Organisms, artifacts, and organizations all evolve and coevolve on rugged, deforming, fitness landscapes. Organisms, artifacts, and organizations, when complex, all face conflicting constraints. So it can be no surprise if attempts to evolve toward good compromise solutions and designs must seek peaks on rugged landscapes. Nor, since the space of possibilities is typically vast, can it be a surprise that even human agents must search more or less blindly. Chess, after all, is a finite game, yet no grand master can sit at the board after two moves and concede defeat because the ultimate checkmate by the opponent 130 moves later can now be seen as inevitable. And chess is simple compared with most of real life. We may have our intentions, but we remain blind watchmakers. We are all, cells and CEOs, rather blindly climbing deforming fitness landscapes. If so, then the problems confronted by an organization—cellular, organismic, business, governmental, or otherwise—living in niches created by other organizations, is preeminently how to evolve on its deforming landscape, to track the moving peaks.

Tracking peaks on deforming landscapes is central to survival. Landscapes, in short, are part of the search for excellence—the best compromises we can attain.

The Logic of Patches

I intend, in this chapter, to describe some recent work I am carrying out with Bill Macready and Emily Dickinson. The results hint at something deep and simple about why flatter, decentralized organizations may function well: contrary to intuition, breaking an organization into "patches" where each patch attempts to optimize for its own selfish benefit, even if that is harmful to the whole, can lead, as if by an invisible hand, to the welfare of the whole organization. The trick, as we shall see, lies in how the patches are chosen. We will find an ordered regime where poor compromises for the entire organization are found, a chaotic regime where no solution is ever agreed on, and a phase transition between order and chaos where excellent solutions are found rapidly. We will be exploring the logic of patches. Given the results I will describe, I find myself wondering whether these new insights will help us understand how complex organizations evolve, and perhaps even why democracy is such a good political mechanism to reach compromises between the conflicting aspirations of its citizens.

The work is all based on our now familiar friend, the *NK* model of rugged fitness landscapes. Since this is so, caveats are again in order. The *NK* model is but one family of rugged, conflict-laden, fitness landscapes. Any efforts here require careful extension. For example, we will need to be far surer than I now am that the results I will discuss extend to other conflict-laden problems, ranging from the design of complex artifacts such as aircraft, to manufacturing facilities, organizational structures, and political systems.

NK fitness landscapes are examples of what mathematicians call hard combinatorial optimization problems. In the framework of *NK* landscapes, the optimization problem is to find either the global optimum, the highest peak, or at least excellent peaks. In *NK* landscapes, genotypes are combinatorial objects, composed of *N* genes with either 1 or 0 allele states. Thus as *N* increases, one finds what is called a combinatorial explosion of possible genotypes, for the number of genotypes is 2^N. One of these genotypes is the global peak we seek. So as *N* goes up, finding the peak can become very much harder. Recall that for $K = N - 1$, the maximum density of interconnectedness, landscapes are fully random and the number of local peaks is $2^N/(N + 1)$. In Chapter 8, we discussed finding maximally compressed algorithms to perform a computation, and noted that such algorithms "live" on random landscapes. Therefore, finding a maximally compressed program for an algorithm amounted to finding one or, at most, a very few of the highest peaks on such a random landscape. Recall that on random landscapes, local hill-climbing soon becomes trapped on local peaks far from the global optimum. Therefore, finding the global peak or one of a few excellent peaks is a completely intractable problem. One would have to search the entire space to be sure of success. Such problems are known as *NP*-hard. This means roughly that the search time to solve the problem increases in proportion to the size of the problem space, which itself increases exponentially because of the combinatorial explosion.

Evolution is a search procedure on rugged fixed or deforming landscapes. No search procedure can guarantee locating the global peak in an *NP*-hard problem in less time than that required to search the entire space of possibilities. And that, as we have repeatedly seen, can be hyperastronomical. Real cells, organisms, ecosystems, and, I suspect, real complex artifacts and real organizations never find the global optima of their fixed or deforming landscapes. The real task is to search out the excellent peaks and track them as the landscape deforms. Our "patches" logic appears to be one way complex systems and organizations can accomplish this.

Before discussing patch logic, I am going to tell you about a well-known procedure to find good fitness peaks. It is called simulated annealing, and was invented by Scott Kirkpatrick at IBM and his colleagues some years ago. The test example of a hard combinatorial optimization problem they chose is the famous Traveling Salesman problem. If one could solve this, one would break the back of many hard optimization problems. Here it is: you, a salesperson, live in Lincoln, Nebraska. You must visit 27 towns in Nebraska, one after the other, and then return home. The catch is, you are supposed to travel your circuit by the shortest route possible.

That's all there is to it. Just get in your flivver, pack 27 lunches in your carry-along ice chest, and get on with the tour. Simple as pie. Or so it sounds.

If the number of cities, N, which is 27 here, increases to 100 or 1,000, the complexity of the pie becomes one of those hyperastronomical kinds of messes. You see, if you start in Lincoln, you have to choose the first city you will go to, and you have 26 choices. After picking the first city on your tour, you must pick the second, so 25 choices remain. And so on. The number of possible tours starting from Lincoln is $(27 \times 26 \times 25 \times \ldots 3 \times 2 \times 1)/2$. (Divide by 2 because you could choose either of two routes around any circuit, and dividing by 2 keeps us from counting the same tour twice.)

One would think there was some easy way to find the very best tour; however, this appears to be a dashed hope. It appears that in order to find the shortest tour, you have to consider them all. As the number of cities, N, grows large, you get one of the combinatorial explosions of possibilities that make genotype spaces and other combinatorial spaces so huge. Even with the fastest computer, it can be impossible to guarantee finding the shortest tour in, say, the life of the human species or the universe.

The best thing to do—indeed, the only practical thing to do—is to choose a route that is excellent, but not necessarily the very best. As in all of life, the salesperson in search of excellence will have to settle for less than perfection.

How can one find at least an excellent tour? Kirkpatrick and his colleagues offered a powerful approach in their concept of simulated annealing. First we need a fitness, or cost, landscape. Then we will search across that landscape to find good short tours. The landscape in question is simple. Consider our 27 cities and all possible tours through them. As we have seen, there is a very large number of such tours. Now we need the idea of which tours are "near" each other, just as, for genotypes, we needed the idea of near mutant neighbors. One way of doing this is to think of a "swap" that exchanges the positions of two cities in a tour. Thus we might have gone A B C D E F A. If we exchange C and F, we would go A B F D E C A (Figure 11.1).

Once we have defined this notion of a "neighboring tour," we can arrange all the possible tours in a high-dimensional space, rather like a genotype space, in which each tour is next to all its neighbors. It's hard to show the proper high-dimensional tour space. Recall that in the *NK* model, each genotype, such as (1111), is a vertex in a Boolean hypercube and is next to N other genotypes: (0111), (1011), (1101), and (1110). Adaptive walks occur from genotype to neighboring genotype

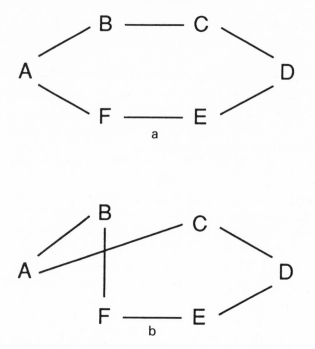

Figure 11.1 *The Traveling Salesman problem. One tries to find the shortest route through several cities. (a) Traveling Salesman tour through six cities, A–F. (b) Exchanging two cities from (a) alters the tour to a "neighboring" tour.*

until a local peak is reached. In tour space, each tour is a "vertex" and is connected by a line to each of its neighboring tours. Since we are seeking the shortest tour from Lincoln through the 27 cities and back to Lincoln, it makes sense to think of the length of the tour as its "cost." Since each tour has a cost, we get a cost landscape over our tour space. Since we are trying to minimize cost rather than maximize fitness, on a cost landscape we seek the deepest valleys rather than the highest peaks. The idea, however, is obviously the same.

Like any other rugged landscape, the tour space may be correlated in a variety of ways; that is, neighboring tours will tend to have the same length, and hence cost. If so, it would be smart to use the correlations to help find excellent tours, even if we cannot find the best. Recall from Chapter 8 that we found a rather general feature of many rugged landscapes: the deepest valleys "drain" the largest region of the space of possibilities. If we think of the valleys as real valleys in a mountainous terrain, water can flow downhill to the deepest valleys from the greatest number of initial positions on the landscape. This feature is essential to simulated annealing, as we are about to see.

Picture a water droplet or a ball. Once it reaches one local minimum, it is stuck there forever unless disturbed by some outside process. That is, whether hiking uphill toward fitness peaks or downhill toward cost minima, if one can take only steps that improve the situation, one will soon become trapped. But the minimum or maximum by which one is trapped may be very poor compared with the excellent minima or maxima. Thus the next question is how to escape.

Real physical systems have an utterly natural way to escape from bad local minima. Sometimes they randomly move in the wrong direction, taking a step uphill when it would seem natural to go down. This random motion is caused by thermal vibration and can be measured by temperature.

Think of a system of interacting molecules colliding with one another. The rate of collisions depends on the velocities of the molecules. Temperature is a measure of the average motion of the molecules, the average kinetic energy. High temperature means that the molecules are in violent random motions. Zero temperature means the molecules are not moving at all.

At a high temperature, a physical system jostles around in its space of possible configurations, molecules colliding with one another and exchanging kinetic energy. This jostling means that the system does not just flow downhill into local energy minima, but can, with a probability that increases with the temperature, jump uphill in energy over "energy barriers" into the drainage basins of neighboring energy minima. If the temperature were lower, the system would be less likely to jump uphill in energy over any given barrier, and hence would be more likely to remain in any given energy "well."

Annealing is just gradual cooling. Real physical annealing corresponds to taking a system and gradually lowering its temperature. A smithy hammering red-hot iron, repeatedly plunging the forming object into cold water and then reheating it and hammering it again, is practicing real annealing. As the smithy anneals and hammers, the microscopic arrangements of the atoms of iron are rearranged, giving up poor, relatively unstable, local minima and settling into lower-energy minima corresponding to harder, stronger metal. As the repeated heating and hammering occurs, the microscopic arrangements in the worked iron can at first wander all over the space of configurations, jumping over energy barriers between all local energy minima. As the temperature is lowered, it becomes harder and harder to jump over these barriers. Now comes the crux assumption: if the deepest energy minima drain the biggest basins, then as temperature is lowered, the microscopic arrangements will tend to become trapped in the biggest drainage basins, precisely because they are the biggest, and drift down to the deepest, stablest energy minima. Working iron by real annealing will achieve hard, strong metal because annealing drives the microscopic atomic arrangements to deep energy minima.

Simulated annealing uses the same principle. In the case of the Traveling Salesman problem, one moves from a tour to a neighboring tour if the second tour is shorter. But, with some probability, one also accepts a move in the wrong direction—to a neighboring tour that is longer and "costs" more. The "temperature" of the system specifies the probability of accepting a move that increases cost by any given amount. In the simulation, the algorithm wanders all over the space of possible tours. Lowering temperature amounts to decreasing the probability of accepting moves that go the wrong way. Gradually, the algorithm settles into the drainage basins of deep, excellent minima.

Simulated annealing is a very interesting procedure to find solutions to conflict-laden problems. In fact, currently it is one of the best procedures known. But there are some important limitations. First, finding good solutions requires "cooling" very slowly. It takes a long time to find very good minima. There is a second obvious problem with anything like simulated annealing if one also has in mind how human agents or organizations might find good solutions to problems in real life. Consider a fighter pilot streaking toward battle. The situation is fast-paced, intense, life-threatening. The pilot must decide on tactics that optimize chances for success in the conflict-laden situation. Our fighter pilot is unlikely to be persuaded to choose his tactics in the heat of battle by making what he knows are very many mistakes with ever-decreasing frequency until he settles into a good strategy. Nor, it would appear, do human organizations try to optimize in anything like this fashion. Simulated annealing may be a superb procedure to find excellent solutions to hard problems, but in real life, we never use it. We simply do not spend our time making mistakes on purpose and lowering the frequency of mistakes. We all try our best, but fail a lot of the time.

Have we evolved some other procedure that works well? I suspect we have, and call it by a variety of names, from federalism to profit centers to restructuring to checks and balances to political action committees. Here I'll call it patches.

The Patch Procedure

The basic idea of the patch procedure is simple: take a hard, conflict-laden task in which many parts interact, and divide it into a quilt of nonoverlapping patches. Try to optimize within each patch. As this occurs, the couplings between parts in two patches across patch boundaries will mean that finding a "good" solution in one patch will change the problem to be solved by the parts in the adjacent patches. Since changes in each patch will alter the problems confronted by the neighboring patches, and the adaptive moves by those patches in turn will alter the problem faced by yet other patches, the system is just like our model coevolving ecosystems. Each patch is the analogue of what we called a species in Chapter 10. Each patch climbs toward fitness peaks on its own landscape, but in doing so deforms the fitness landscapes of its partners. As we saw, this process may spin out of control in Red Queen chaotic behavior and never converge on any good overall solution. Here, in this chaotic regime, our system is a crazy quilt of ceaseless changes. Alternatively, in the analogue of the evolutionary stable strategy (ESS) ordered regime, our system might freeze up, getting stuck on poor local peaks. Ecosystems, we saw, attained the highest average fitness if poised between Red Queen chaos and ESS order. We are about to see that if the entire conflict-laden task is broken into the properly chosen patches, the coevolving system lies at a phase transition between order and chaos and rapidly finds very good solutions. Patches, in short, may be a fundamental process we have evolved in our social systems, and perhaps elsewhere, to solve very hard problems.

By now you know the *NK* model. All it consists of is a system of *N* parts, each of which makes a "fitness contribution" that depends on its own state and the states of *K* other parts. Let's put the *NK* model onto a square lattice (Figure 11.2). Here each part is located at a vertex connecting it to its four immediate neighbors: north, south, east, and west. As before, we let each part have two states: 1 and 0. Each part makes a fitness contribution depending on its own state and that of its north, south, east, and west neighbors. This fitness contribution is assigned at random between 0.0 and 1.0. As before, we can define the fitness of the entire lattice as the average of the fitness contribution of each of the parts. Say all are in the 1 state. Just add up the contributions of each part and divide by the number of parts. Do this for all possible configurations and we get a fitness landscape.

Macready, Dickinson, and I have been looking at fairly large lattices, 120 × 120, so our model hard problems have 14,400 parts. That should be enough to be hard. Since *NK* rugged landscapes are very similar to the landscapes of many hard conflict-laden optimization problems, including the Traveling Salesman, finding means to achieve good optima here is likely to be generally useful. Notice, as usual, the vast space of possibilities. Since each part can be in two states, 1 and 0, the total number of combinations of states of the parts, or configurations of the

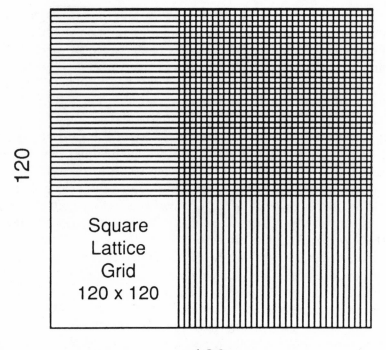

120

Square
Lattice
Grid
120 x 120

120

Figure 11.2 *An* NK *model in the form of a 120 × 120 square lattice. Each site, which can be in one of two states, 1 or 0, is connected to its four neighbors: north, south, east, and west. (The lattice is bent into the surface of torus; that is, the top edge is "glued" to the bottom edge, and the left edge is glued to the right edge so that each site has four neighbors.)*

lattice, is $2^{14,400}$. Forget it, there is not enough time since the Big Bang to find the global optimum. We seek excellence, not perfection.

Because Bill Macready is a physicist and physicists like to minimize an "energy" rather than maximize a "fitness," and because we are also used to minimizing on cost surfaces, let us think of the *NK* model as yielding an "energy" landscape and minimize energy. Each configuration of the 14,400 parts on the 120 × 120 lattice is a vertex on the 14,400-dimensional Boolean hypercube. Each vertex is an immediate neighbor of 14,400 other vertices, each corresponding to flipping one of the 14,400 parts to the other state, 1 or 0. Each vertex has an energy, so the *NK* model generates an energy landscape over this huge Boolean hypercube of configurations. We seek the deep, excellent minima. The *NK* landscape remains unchanged here, we just go "downhill" rather than "uphill" on it. The statistical-landscape features are the same in either direction.

Now I will introduce patches. Suppose we use the same *NK* land-scape, leave the parts coupled in the same ways, but divide the system into nonoverlapping quilt patches of different sizes. The rule is always going to be the same: try flipping a part to the opposite value, 1 to 0 or 0 to 1. If this move lowers the energy of the patch in which the part is located, accept the move; otherwise, do not accept the move.

Figure 11.3 shows a smaller version of our 120 × 120 lattice, here reduced to a 10 × 10 square. In Figure 11.3*a*, we consider the entire lattice as a single "whole" patch. I'll call this the "Stalinist" limit. Here a part can flip from 1 to 0 or 0 to 1 only if the move is "good" for the en-

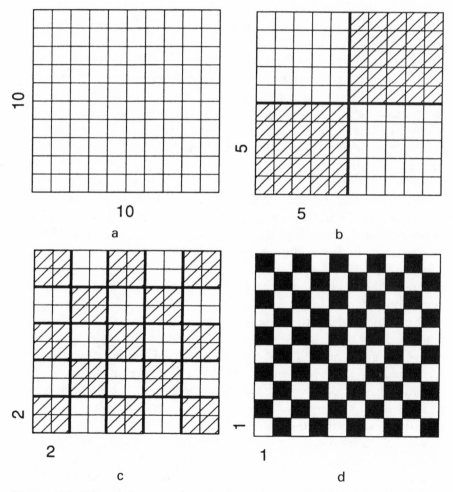

Figure 11.3 *The patches procedure.* (a) *A 10 × 10 NK lattice; the entire system is one patch.* (b) *The same lattice divided into four 5 × 5 patches,* (c) *25 2 × 2 patches, and* (d) *100 1 × 1 patches.*

tire lattice, lowering its energy. We all must act for the benefit of the entire "state."

Since in the Stalinist limit any move must lower the energy of the entire lattice, then as successive parts are tried and some are flipped, the entire system will carry out an adaptive walk to some local energy minimum and then stay there forever. Once at such a local minimum, no part can be flipped and find a lower energy for the total lattice, so no flips will be accepted. All parts "freeze" into an unchanging state, 1 or 0. In short, in the Stalinist regime, the system locks into some solution and is frozen there forever. The Stalinist regime, where the game is one for all, one for the state, ends up in frozen rigidity.

Now look at Figure 11.3*b*. Here the same square lattice, with the same couplings among the parts, is broken into four patches, each a 5 × 5 sublattice of the entire 10 × 10 lattice. Each part belongs to only a single patch. But the parts near the boundaries of a patch are coupled to parts in the adjacent patches. So adaptive moves, parts flipping between 1 and 0 states, in one patch will affect the neighboring patches. I emphasize that the couplings among the parts are the same as in the Stalinist regime. But now, by our rule that a part can flip if it is good for the patch in which it is located, a part might help its own patch, but hurt an adjacent patch.

In the Stalinist limit, the entire lattice can flow only downhill toward energy minima. Thus the system is said to flow on a potential surface. The system is like the ball on a real surface in a valley. The ball will roll to the bottom of the valley and remain there. Once the lattice is broken into patches, however, the total system no longer flows on a potential surface. A flip of a part in one patch may lower the energy of that patch, but *raise* the energies of adjacent patches because of the couplings across boundaries. Because adjacent patches can go up in energy, the total energy of the lattice itself can go up, not down, when a single patch makes a move to lower its own energy. And since the entire lattice can go up in energy, the total system is not evolving on a potential surface. Breaking the system into patches is a little like introducing a temperature in simulated annealing. Once the system is broken into quilt patches so an adaptive move by one patch can "harm" the whole system, that move causes the whole system to "go the wrong way."

We reach a simple, but essential conclusion: once the total problem is broken into patches, the patches are coevolving with one another. An adaptive move by one patch changes the "fitness" and deforms the fitness landscape, or, alternatively, the "energy" and the "energy landscape," of adjacent patches.

It is the very fact that patches coevolve with one another that begins to hint at powerful advantages of patches compared with the Stalinist

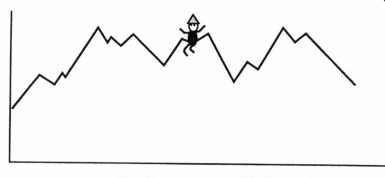

Position on Landscape

Figure 11.4 *An energy landscape showing a system trapped on a poor, high-energy local minimum.*

limit of a single large patch. What if, in the Stalinist limit, the entire lattice settles into a "bad" local minimum, one with high energy rather than an excellent low-energy minimum, as in Figure 11.4? The single-patch Stalinist system is stuck forever in the bad minimum. Now let's think a bit. If we break the lattice up into four 5 × 5 patches just after the Stalinist system hits this bad minimum, what is the chance that this bad minimum is not only a local minimum for the lattice as a whole, but also a local minimum for each of the four 5 × 5 patches individually? You see, in order for the system broken into four patches to "stay" at the same bad minimum, it would have to be the case that the same minimum of the entire lattice happens also to be a minimum for all four of the 5 × 5 patches individually. If not, one or more of the patches will be able to flip a part, and hence begin to move. Once one patch begins to move, the entire lattice is no longer frozen in the bad local minimum.

Well, the intuitive answer is pretty obvious. If the entire lattice, in the Stalinist limit, flows to a bad local minimum, the chance is small that the same configuration of the parts is also a local minimum for all four of the 5 × 5 patches, so the system will not remain frozen. It will "slip" away and be able to explore further across the total space of possibilities.

The Edge of Chaos

We have now encountered a phase transition between order and chaos in model genomic networks and in coevolutionary processes. In Chapter 10, we saw that the highest average fitness in coevolving systems appeared to arise just at the phase transition between Red Queen chaos and ESS order. Breaking large systems into patches allows the patches literally to coevolve with one another. Each climbs toward its fitness peaks, or energy minima, but its moves deform the fitness landscape or energy landscape of neighboring patches. Is there an analogue of the chaotic Red Queen regime and the ordered ESS regime in patched systems? Does a phase transition between these regimes occur? And are the best solutions found at or near the phase transition? We are about to see that the answer to all these questions is yes.

The Stalinist limit is in the ordered regime. The total system settles to a local minimum. Thereafter, no part can be flipped from 1 to 0 or 0 to 1. All the parts, therefore, are frozen. But what happens at the other limit? In the extreme situation shown in Figure 11.3d, each part constitutes its own patch. In this limit, on our 10×10 lattice, we have created a kind of "game" with 100 players. At each moment, each player considers the states, 1 or 0, of its north, south, east, and west neighbors and takes the action, 1 or 0, that minimizes its own energy. It is easy to guess that in this limit, call it the "Leftist Italian" limit, the total system never settles down. Parts keep flipping from 1 to 0 and 0 to 1. The system is in a powerfully disordered, or chaotic, regime.

Since the parts never converge onto a solution where they stop flipping, the total system burbles along with quite high energy. In the NK model, the expected energy of a randomly chosen configuration of the N parts without trying to minimize anything is 0.5. (The 0.5 value follows because we assigned fitness or energy as random decimals between 0.0 and 1.0, and the average value is halfway between these limits, or 0.5.) In the chaotic Leftist Italian limit, the average energy achieved by the lattice is only a slight bit less, about 0.47. In short, if the patches are too numerous and too small, the total system is in a disordered, chaotic regime. Parts keep flipping between their states, and the average energy of the lattice is high.

We come to the fundamental question: At what patch size does the overall lattice actually minimize its energy?

The answer depends on how rugged the landscape is. Our results suggest that if K is low so the landscape is highly correlated and quite smooth, the best results are found in the Stalinist limit. For simple problems with few conflicting constraints, there are few local minima in which to get trapped. But as the landscape becomes more rugged, reflecting the fact that the underlying number of conflicting constraints is becoming more severe, it appears best to break the total system into a number of patches such that the system is near the phase transition between order and chaos.

In the current context, we can look at increasing the number of conflicting constraints by retaining our square-lattice configuration, but considering cases in which each part is affected not only by itself and its four immediate neighbors—north, south, east, and west—but by its eight nearest neighbors—the first four, plus the northwest, northeast, southeast, and southwest neighbors. Figure 11.5 shows that increasing the range across the square lattice over which parts influence one another allows us to think of the four-neighbor, eight-neighbor, 12-neighbor, and 24-neighbor cases. Hence in terms of the *NK* model, *K* is increasing from four to 24.

Figure 11.6 shows the results obtained, allowing the patched lattice to evolve forward in time, flipping randomly chosen parts if and only if the move lowers the energy of the patch containing the part. We carried

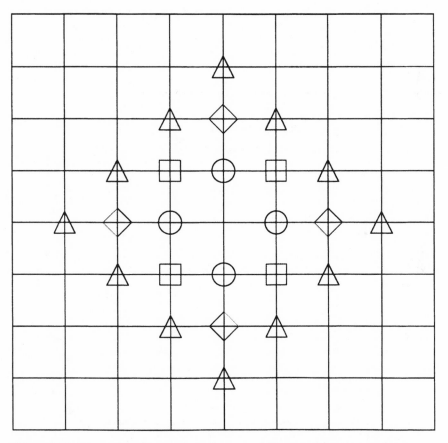

Figure 11.5 *Increasing the range. A site on an* NK *square lattice may be connected to* K *= 4 neighbors (circles); or may "reach" farther out into the square lattice to be coupled to* K *= 8 sites (circles plus squares); or may reach still farther out to be coupled to* K *= 12 sites (circles plus squares plus diamonds); or may reach out to be coupled to* K *= 24 sites (circles plus squares plus diamonds plus triangles).*

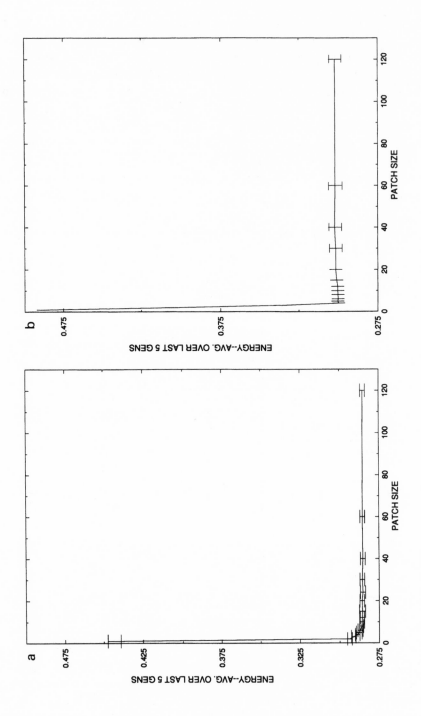

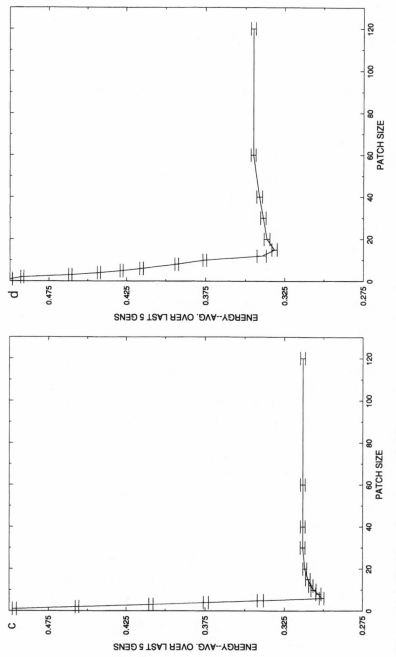

Figure 11.6 *Divide and conquer. As landscapes become more rugged, energy minimums can be more easily found by dividing the problem into patches. Patch sizes are plotted on the x-axis. The average low energy reached at the end of the simulation runs is plotted on the y-axis. (Bars above and below the curve at each patch size correspond to plus and minus 1 standard deviation.) (a) For a smooth, K = 4, landscape, the optimum patch size, minimizing energy, is a 120 × 120 single patch—the "Stalinist" limit. (b) For a more rugged, K = 8, landscape, the optimum patch size is 4 × 4. (c) For K = 12, the optimum patch size is 6 × 6. (d) For K = 24, the optimum patch size is 15 × 15.*

out these simulations as the total energy of the lattice decreased until the energy attained was either fixed or only fluctuating in a narrow range. Figures 11.6a–d plot energy on the y-axis against the size of each of the patches on the x-axis. Here all patches are themselves square, and our results are from the 120 × 120 lattice. All our simulations were carried out on exactly the same set of NK lattices. Only the sizes of the patches were changed. Thus the results tell us the effect of patch size and number on how well the lattice as a whole lowered its overall energy.

The results are clear. When $K = 4$, it is best to have a single large patch. In worlds that are not too complex, when landscapes are smooth, Stalinism works. But as landscapes become more rugged, $K = 8$ to $K = 24$, it is obvious that the lowest energy occurs if the total lattice is broken into some number of patches.

Here, then, is the first main and new result. It is by no means obvious that the lowest total energy of the lattice will be achieved if the lattice is broken into quilt patches, each of which tries to lower its own energy regardless of the effects on surrounding patches. Yet this is true. It can be a very good idea, if a problem is complex and full of conflicting constraints, to break it into patches, and let each patch try to optimize, such that all patches coevolve with one another.

Here we have another invisible hand in operation. When the system is broken into well-chosen patches, each adapts for its own selfish benefit, yet the joint effect is to achieve very good, low-energy minima for the whole lattice of patches. No central administrator coordinates behavior. Properly chosen patches, each acting selfishly, achieve the coordination.

But what, if anything, characterizes the optimum patch-size distribution? The edge of chaos. Small patches lead to chaos; large patches freeze into poor compromises. When an intermediate optimum patch size exists, it is typically very close to a transition between the ordered and the chaotic regime.

In Figure 11.7, I show something of this "phase transition." These two figures concern the same NK landscape on the same lattice, begun from the same initial state. The only difference is that in Figure 11.7a the 120 × 120 lattice was broken into 5 × 5 patches, while in Figure 11.7b the lattice was broken into 6 × 6 patches. The figures show how often each site on the lattice flips. Sites that flip often are dark; ones that do not flip at all are white. In Figure 11.7a, most sites are dark—indeed, darkest along the boundaries between patches. Sites keep flipping in a chaotic, disordered regime. But just increase patch size to 6 × 6 and the results are startling. As shown in Figure 11.7b, almost all sites stop flip-

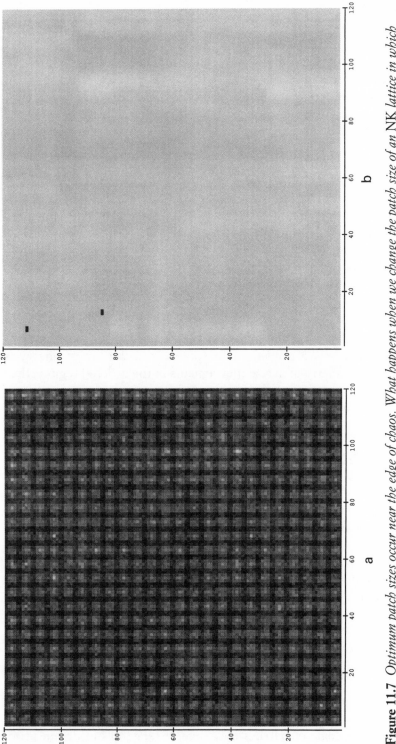

Figure 11.7 *Optimum patch sizes occur near the edge of chaos. What happens when we change the patch size of an NK lattice in which K = 12? (Sites that flip often are dark; sites that do not flip are white.) (a) When broken into 5 × 5 patches, the lattice is in the unfrozen, chaotic regime. Most sites are dark and darkest along the boundaries between patches. (b) But watch what happens when we break the lattice into 6 × 6 patches. Almost all sites stop flipping and freeze. The transition from (a) to (b) is the phase transition from chaos to order.*

ping. A few quarrelsome sites across a few boundaries keep flipping, but almost all sites have settled down and stopped changing.

The system, if broken into 5×5 patches, never converges onto a solution and the energy of the whole lattice is high. The same system, if broken into 6×6 patches, converges onto a solution in which almost all patches and almost all their parts are no longer changing. And the energy of the entire lattice is very low. A kind of phase transition has occurred as the same lattice is broken into 5×5 and then into 6×6 patches, passing from a chaotic to an ordered regime.

In terms of a coevolutionary system, it appears that if the lattice is broken into 6×6 patches, each patch reaches a local minimum that is consistent with the minima of the patches adjacent to it. This global behavior is now like a Nash equilibrium among the patches, or an evolutionary stable strategy. The optimum found by each patch is consistent with the optima found by its neighbors. No patch has an incentive to change further. The patches therefore stop coevolving across their landscapes. The system converges on a total solution.

Across lots of simulations, it appears to be the case that the lowest energy is found for a given lattice in the ordered, ESS regime, somewhere near the phase transition. In some cases, the lowest energy is found at the smallest patch size that remains in the ordered regime, thus just before the phase transition to chaos. In other cases, the lowest energy is found when patch size is still a bit larger so the system is a bit deeper into the ordered regime and farther from the phase transition to chaos. Therefore, as a general summary, it appears that the invisible hand finds the best solution if the coevolving system of patches is in the ordered regime rather near the transition to chaos.

Patch Possibilities

I find it fascinating that hard problems with many linked variables and loads of conflicting constraints can be well solved by breaking the entire problem into nonoverlapping domains. Further, it is fascinating that as the conflicting constraints become worse, patches become ever more helpful.

While these results are new and require extension, I suspect that patching will, in fact, prove to be a powerful means to solve hard problems. In fact, I suspect that analogues of patches, systems having various kinds of local autonomy, may be a fundamental mechanism underlying adaptive evolution in ecosystems, economic systems, and cultural systems. If so, the logic of patches may suggest new tools in design problems. Moreover, it may suggest new tools in the management of complex organizations and in the evolution of complex institutions worldwide.

Homo sapiens sapiens, wise man, has come a long way since bifacial stone axes. We are constructing global communication networks, and whipping off into space in fancy tin cans powered by Newton's third law. The *Challenger* disaster, brownouts, the Hubble trouble, the hazards of failure in vast linked computer networks—our design marvels press against complexity boundaries we do not understand. I wonder how general it has become as we approach the year 2000 that the design of complex artifacts is plagued with nearly unsolvable conflicting constraints. One hears tales, for example, of attempts to optimize the design of complex manufactured artifacts such as supersonic transports. One team optimizes airfoil characteristics, another team optimizes seating, another works on hydraulics, but the multiple solutions do not converge to a single compromise that adequately solves all the design requirements. Proposals keep evolving chaotically. Eventually, one team makes a choice—say, how the hydraulic system or the airfoil structure will be constructed—and the rest of the design becomes frozen into place because of this choice.

Does this general problem of nonconvergence reflect "patching" the design problem into too many tiny patches such that the overall design process is in a nonconverging chaotic regime, just as would be our 120 × 120 lattice broken into 5 × 5 rather than 6 × 6 patches? If one did not know that increasing patch size would lead from chaos to ordered convergence on excellent solutions, one would not know to try "chunking" bigger. It seems worth trying on a variety of real-world problems.

Understanding optimal patching may be useful in other areas in the management of complex organizations. For example, manufacturing has long used fixed facilities of interlinked production processes leading to a single end product. Assembly-line production of manufactured products such as automobiles is an example. Such fixed facilities are used for long production runs. It is now becoming important to shift to flexible manufacturing. Here the idea is to be able to specify a diversity of end products, reconfigure the production facilities rapidly and cheaply, and thus be able to carry out short production runs to yield small quantities of specialized products for a variety of niche markets. But one must test the output for quality and reliability. How should this be done? At the level of each individual production step? At the level of output of the entire system? Or at some intermediate level of chunking? I find myself thinking that there may be an optimal way to break the total production process in each case into local patches, each with a modest number of linked production steps; optimize within each patch; let the patches coevolve; and rapidly attain excellent overall performance.

Patching systems so that they are poised on the edge of chaos may be extremely useful for two quite different reasons: not only do such systems rapidly attain good compromise solutions, but even more essentially, such poised systems should track the moving peaks on a changing landscape very well. The poised, edge-of-chaos systems are "nearly melted." Suppose that the total landscape changes because external conditions alter. Then the detailed locations of local peaks will shift. A rigid system deep in the ordered regime will tend to cling stubbornly to its peaks. Poised systems should track shifting peaks more fluidly.

Peter Banks urged us to find new insights into how to manage complex organizations. We need, he said, a theory of decentralization that works. If we now begin to have evidence that when properly parsed into the right-size patches, complex problems can rapidly be brought to fruitful compromise solutions, then it appears foolish not to attempt to develop this beginning insight into a rational management technique.

Yet if we are going to try to develop "patches" into a rational management technique, whether in business or more broadly, then we must confront directly the fact that we almost always misspecify the problems we wish to solve. We then solve the wrong problem and stand in danger of applying our solution to the real-world problem we confront.

Misspecification arises all the time. Physicists and biologists, trying to figure out how complex biopolymers such as proteins fold their linear sequence of amino acids into compact three-dimensional structures, build models of the landscape guiding such folding and solve for the deep energy minima. Having done so, the scientists find that the real protein does not look like the predicted one. The physicists and biologists have "guessed" the wrong potential surface; they have guessed the wrong landscape and hence have solved the wrong hard problem. They are not fools, for we do not know the right problem.

Misspecification is, therefore, endemic. Suppose we consider some production facility with interlinked inputs and outputs. An example might be a complex chemical-processing facility in the petroleum industry that cracks oil and builds the smaller molecules into various larger products. We might wish to optimize output performance subject to a variety of conflicting constraints and goals. However, we may be ignorant of some of the details of the chemical reaction rates occurring in the sequence of reactions and their sensitivities to temperature, pressure, and catalyst purity. If we do not know these things, any specific model of the facility we build will almost certainly be misspecified. We might build a computer model of the web of production processes and try patching these in various ways until we find patterns that leave the model system just in the ordered regime. Then we might rush out to the real production facility and attempt to apply the "optimal solution" suggested by our computer model. My bet is that the suggested solution about how to optimize within and between patches to optimize overall performance would be a bust. Because we almost certainly will have misspecified our model in the first place, our optimal answer solves a misstated problem.

We must learn how to learn in the face of persistent misspecification. Suppose we model the production facility, and learn from that model that a particular way to break it into patches is optimal, allowing the system to converge on a suggested solution. If we have misspecified the problem, the detailed solution is probably of little value. But it may often be the case that the optimal way to break the problem into patches is itself very insensitive to misspecifications of the problem. In the *NK* lattice and patch model we have studied, a slight change in the *NK* landscape energies will shift the locations of the minima substantially, but may not alter the fact that the lattice should still be broken into 6 × 6 patches. Therefore, rather than taking the suggested *solution* to the misspecified problem and imposing it on the real facility, it might be far smarter to take the suggested *optimal patching* of the misspecified problem, impose that on the real production facility, and then try to optimize performance within each of the now well-defined patches. In short, learning how to optimize the misspecified problem may not give us the solution to the real problem, but may teach us how learn about the real problem, how to break it into quilt patches that coevolve to find excellent solutions.

Receiver-Based Optimization: Ignoring Some of the "Customers" Some of the Time

Larry Wood is a warm and rather brilliant young scientist at GTE who keeps winning awards for sundry wild ideas that he, somehow, pulls off. Wood showed up at the Santa Fe Institute one day in the spring of 1993, very excited about patches. "You guys have got to be thinking about receiver-based communication! I'll help."

It turned out that receiver-based communication is roughly this: all the agents in a system that is trying to coordinate behavior let other agents know what is happening to them. The receivers of this information use it to decide what they are going to do. The receivers base their decisions on some overall specification of "team" goal. This, it is hoped, achieves coordination. Wood explained that the U.S. Air Force had adopted this procedure to allow pilots to coordinate mutual behavior largely in the absence of ground control. The pilots talk to one another and respond preferentially to those nearest them, and achieve collective coordination in a way loosely analogous to flocking behavior in birds.

We have now begun studying a simple version of receiver-based communication in, of course, our beloved *NK* model. The preliminary results are very intriguing.

Let us again put our *NK* "parts," or sites, onto a 120 × 120 square lattice. Think of each site as a "provider" that affects itself and $K = 4$ "customers": north, south, east, and west. At each moment, each customer announces to each of its different providers what will happen to that customer if that provider changes from 1 to 0, or from 0 to 1. Then each customer relies on its providers to make "wise" decisions based on the communications received by that provider.

Thus each site is a separate agent facing a separate optimization problem in which it is to act for the benefit of itself and its four customers. So here is a simple model of receiver-based communication, where the provider-agents are confronted with a very complex, conflict-laden problem by the information received from their customers. Pause to tune intuition. In the patches approach, a site near a patch boundary might have a chance to flip, and will do so for the benefit of its own patch, regardless of the effects on sites in adjacent patches. The site ignores some of its "customers." This allows the total system to "go the wrong way" and slide off bad local minima. But only sites near boundaries can do this in patches, for sites in the middle of large patches do not affect customers in other patches.

This observation suggests that it might be useful if, in our receiver-based communication system, we allowed sites to ignore some of their customers. Let's say that each site pays attention to itself and a fraction, *P*, of its customers, and ignores $1 - P$ of them. What happens if we "tune" *P*?

What happens is shown in Figure 11.8. The lowest energy for the entire lattice occurs when a small fraction of customers is ignored! As Figure 11.8 shows, if each site tries to help itself and all its customers, the system does less well than if each site pays attention, on average, to about 95 percent of its customers. In the actual numerical simulation, we do this by having each site consider each of its customers and pay attention to that customer with a 95 percent probability. In the limit

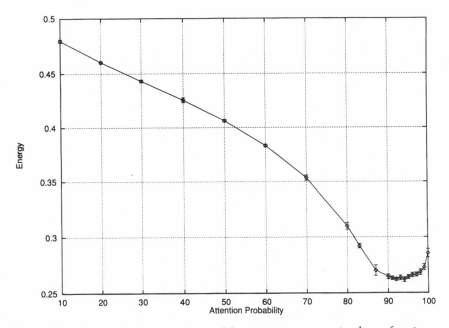

Figure 11.8 *You can't please everyone. Minimum energy attained as a fraction, P, of "customer" sites to which each provider site pays attention. The minimum is attained when about 93 percent to 95 percent of the customer sites are attended to at each moment, while 5 percent to 7 percent of the customer sites are ignored.*

where each site pays attention to no customers, of course, energy of the entire lattice is very high, and hence bad.

So we learn something interesting. In hard conflict-laden problems, the best solutions may be found if, in some way, different subsets of the constraints are ignored at different moments. You should not try to please all of the people all of the time, but you should pay attention to everyone some of the time! Hmmm, sounds familiar doesn't it? It sounds like real life.

Although simulated annealing does better than patches in our test cases, it does little to deepen our understanding of how individuals and organizations solve hard problems. Human organizations do divide themselves into departments, profit centers, and other decentralized structures. Further, receiver-based optimization appears to do about as well as simulated annealing. Since humans do not make mistakes on purpose and make them with decreasing frequency, but humans do ignore constraints some of the time out of ignorance or otherwise, I suspect that patching and receiver-based optimization are part of how we really try to coordinate complex, conflict-laden tasks.

Patch Politics

Might there be implications of patches and receiver-based optimization for politics? The intuition seems strong that there may be a fruitful connection. Democracy is sometimes considered, simplistically, as "rule of the majority." Of course, democracy is a very much more complicated governmental process. The United States Constitution and Bill of Rights create a federal system, where the whole is broken into parts, or patches, called states. States are themselves composed of patches—counties, municipalities. Jurisdictions often overlap. Individual rights, those of the smallest patches, are guaranteed.

The polity in any country presents an enormous variety of conflicting desires, requirements, demands, interests. Might we best think of democracy as a system to achieve excellent compromises among these conflicting interests? That point is almost a truism. What has not been obvious, however, is the possibility that the apparently disjointed, opportunistic, fractured, argumentative, pork-barrel, swap, cheat, steal-a-vote, cluttered system may actually work pretty well because it is a well-evolved system to solve hard, conflict-laden problems and find, on average, pretty good compromises.

Aquinas attempted to find a self-consistent moral code. Kant sought the same in his brilliant maxim, Act such that the principle of your action can be universalized. Hence telling the truth makes sense because if all agents do so, then telling the truth is a consistent action. But if everyone lies, there is no point of lying. Lying makes sense only in a system where most agents tell the truth.

Great tries by Aquinas and Kant. Nevertheless, these hopes for consistency stumble in the real world. No one guarantees that the set of goals held as "good" will be mutually consistent. Or constant in time. We necessarily live in and make conflict-laden worlds. Therefore, our political machinery must evolve toward procedures that find good compromises. Patches and receiver-based optimization have a ring of reality and natural plausibility about them.

Furthermore, patches and receiver-based optimization may afford us a rather new conceptual tool to understand the workings of democracy. I am not a political scientist, so I asked the advice of Robert Axelrod, the MacArthur Fellow who did the wonderful work on repeated Prisoner's Dilemma, showing when cooperation might emerge. Recently, Axelrod has been looking at simple models in which new higher-level political "actors" could emerge. His new model is based on states that intimidate their neighbors, demand tribute, and then form cooperating alliances with those tribute states for mutual benefit. The alliances emerge as new actors.

When I asked Axelrod about patches and democracy, he told me that a federal system with partitioning into local semiautonomous regions could be thought of as a mechanism that allowed "experimentation" such that novel solutions might be invented "locally," and then copied elsewhere. Oregon innovates; the rest of the country imitates. This exploratory behavior seems fully plausible, and actually occurs. But patches and receiver-based optimization suggest something more—that a democratic system can find excellent compromises among conflicting constraints. In a Stalinist regime, the system may typically settle determinedly onto a very poor compromise solution and then remain there rigidly. In the fanciful Leftist Italian regime, the system rarely settles down to any solution at all. In the extreme, each citizen is his or her own political action committee. No compromises are settled on.

As mentioned in Chapter 1, James Mill once deduced from what he considered indubitable first principles that a constitutional monarchy remarkably like that of England in his day was obviously the highest form of government. One is always in danger of deducing the optimality of the familiar. Let's call this a Mill-mistake. God knows, we all suffer the danger.

Yet I suspect that understanding how patches and receiver-based optimization help provide mechanisms allowing complex systems to reach excellent compromises may not mislead. We may have stumbled into new grounds to see democracy, stumbling governmental mode of an increasing fraction of the globe, as utterly natural. If so, the evolution of democracy takes its place among the understood phenomena. We, at home in the universe, having made and remade our local patch of it together.

Chapter 12

An Emerging
Global Civilization

Hard rain had started. Brian Goodwin and I dodged under low-hanging brush into a squat rectangular opening in a low concrete structure buried into the crest of a hillside overlooking Lago di Lugano in northern Italy, a few kilometers from the Swiss border. We found ourselves in a World War I bunker peering though a horizontal slot made for machine guns as the rain pelted the lake. We were able to see the imagined route where the hero of Hemingway's *A Farewell to Arms* made his way across the lake to the Swiss shore a scant three kilometers away. I had traversed the same route in a small rented sailboat with my young children, Ethan and Merit, two days before and bought them hot chocolate as fortification for our return crossing. Brian was visiting us at the home of my mother-in-law, Claudia, in Porto Ceresio, set on the lake's edge. He and I figured we would think through the implications of autocatalytic sets and functional organization.

Brian Goodwin is a close friend, a Rhodes Scholar from Montreal years ago, and a wonderful theoretical biologist. I first met him in the office of Warren McCulloch, one of the inventors of cybernetics, at the Research Laboratory of Electronics at MIT in 1967. Brian, and several years later my wife, Elizabeth, and I, had had the privilege of being invited to live with Warren and his wife, Rook, for several months. When he had visited McCulloch, Brian was working on the first mathematical model of large networks of genes interacting with one another to control cell differentiation. I still remember the mixture of awe and dread that crossed my heart when, as a young medical student stumbling forward with my own first efforts at the Boolean network models I have described, I fell upon Goodwin's book, *The Temporal Organization of Cells,* in a bookstore in San Francisco. Every young scientist, at one point or another, faces this moment: "Oh, God, someone has already done it!" Typically, someone hasn't quite done what you have set out to do, so your entire life, about to vanish into an abyss of lost dreams, can find a narrow passage forward toward some upland pasture. Brian had not done quite what I was trying to do, although the spirit was similar. We have been fast friends for years. I deeply admire his sense of the unseen deep issues in biology.

"Autocatalytic sets," mused Brian as we watched rain turn to hail and spatter down, "those autocatalytic sets are absolutely natural models of functional integration. They are functional wholes." Of course, I agreed with him. Some years before, two Chilean scientists, Humberto Maturana, himself a close colleague of McCulloch, and Francisco Varela, had formulated an image of what they called autopoesis. Autopoetic systems are those with the power to generate themselves. The image is older than Maturana, who I later met in India, and Varela, who has become a good friend. Kant, writing in the late eighteenth century, thought of organisms as autopoetic wholes in which each part existed both for and by means of the whole, while the whole existed for and by means of the parts. Goodwin and his colleague Gerry Webster had written a clean exposition of the intellectual lineage leading from Kant to contemporary biology. They had noted that the idea of an organism as an autopoetic whole had been replaced by the idea of the organism as an expression of a "central directing agency." The famous biologist August Weismann, at the end of the nineteenth century, developed the view that the organism is created by the growth and development of specialized cells, the "germ line," which contains microscopic molecular structures, a central directing agency, determining growth and development. In turn, Weismann's molecular structures became the chromosomes, became the genetic code, became the "developmental program" controlling ontogeny. In no small measure, the intellectual lineage is straight from Weismann to today. In this trajectory, we have lost an earlier image of cells and organisms as self-creating wholes. The entire explanatory burden is placed on the "genetic instructions" in DNA—master molecule of life—which in turn is crafted by natural selection. From there it is a short step to the notion of organisms as arbitrary, tinkered-together contraptions. The code can encode any program, and hence any tinkered bric-a-brac that selection may have cobbled together.

Yet, as we saw in Chapter 3, it is not implausible that life emerged as a phase transition to collective autocatalysis once a chemical minestrone, held in a localized region able to sustain adequately high concentrations, became thick enough with molecular diversity. A collectively autocatalytic set of molecules—at least in silico, as we have seen—is capable of reproducing itself, dividing into two "blobs" capable of heritable variation, and hence, following Darwin's argument, capable of evolution. But in a collectively autocatalytic set, there is no central directing agency. There is no separate genome, no DNA. There is a collective molecular autopoetic system that Kant might have been heartened to behold. The parts exist for and by means of the whole; the whole exists for and by means of the parts. Although not yet achieved in a laboratory beaker, an autocatalytic set is not mysterious. It is not yet a true organism. But if we stumbled on some evolving or even coevolving autocatalytic sets in a test tube or hydrothermal vent, we'd tend to feel we were looking at living systems.

Whether or not I am right that life started with collective autocatalysis, the mere fact that such systems are possible should make us question the dogma of a central directing agency. The central directing agency is not necessary to life. Life has, I think, an inalienable wholeness. And always has.

So Brian and I actually had a large agenda in the background as we squatted, wondering at the origins of life and the regularity of war and death spewing out the concrete machine-gun slot before us. We could see that collectively autocatalytic sets of molecules such as RNA or peptides announce their functional wholeness in a clear, nonmystical way. A set of molecules either does or does not have the property of catalytic closure. Catalytic closure means that every molecule in the system either is supplied from the outside as "food" or is itself synthesized by reactions catalyzed by molecular species within the autocatalytic system. Catalytic closure is not mysterious. But it is not a property of any single molecule; it is a property of a system of molecules. It is an emergent property.

Once we have autocatalytic sets, we can see that such systems could form an ecology of competitors and mutualists. What you "squirt" at me may poison or abet some reaction of mine. If we help one another, we may have advantages of trade. We can evolve toward close coupling, symbiosis, and the emergence of higher-ordered entities. We can form a molecular "economy." Ecology and economy are already implicit in coevolving autocatalytic sets. Over time, Brian and I imagined, such an ecology of autocatalytic systems interacting with one another, coevolving, would explore an increasing domain of molecular possibilities, creating a biosphere of expanding molecular diversity in some as yet unclear way. A kind of molecular "wave front" of different kinds of molecules would propagate across the globe.

Later, the same image would begin to appear dimly like a wave front of technological innovations and cultural forms, like an emerging global civilization created by us, the descendants of *Homo habilis,* who may have first wondered at his origins and destiny around some sputtering fire. Perhaps it was on the shore of Lago di Lugano one night. Perhaps hard rain fell.

The rain stopped; we crawled out of the bunker and headed down to Claudia's house, hoping that she and Elizabeth had made polenta and funghi as well as minestrone. We felt that our vision was promising. But we knew we were stuck. We hadn't a clue how to generalize from an image of proteins or RNA molecules acting on one another to a broader framework. We would have to wait six years until Walter Fontana invented a candidate for the broader framework.

Alchemy

Walter Fontana is a young theoretical chemist from Vienna. His thesis work, with Peter Schuster, concerned how RNA molecules fold into complex structures and how evolution of such structures might occur. Fontana and Schuster, like Manfred Eigen, like me, like others, were beginning to consider the structure of molecular fitness landscapes of the type discussed in Chapter 8.

But Fontana harbored more radical aims. Visiting Eigen's group at Göttingen, Fontana found himself in conversation with John McCaskill, an extremely able young physicist engaged in theory and experiments evolving RNA molecules. McCaskill, too, had a more radical aim.

Turing machines are universal computational devices that can operate on input data, which can be written in the form of binary sequences. Referring to its program, the Turing machine will operate on the input tape and rewrite it in a certain way. Suppose the input consisted of a string of numbers, and the machine was programmed to find their average value. By changing 1 and 0 symbols on the tape, the machine will convert it into the proper output. Since the Turing machine and its program can *themselves* be specified by a sequence of binary digits, one string of symbols is essentially manipulating another string. Thus the operation of a Turing machine on an input tape is a bit like the operation of an enzyme on a substrate, snipping a few atoms out, adding a few atoms here and there.

What would happen, McCaskill had wondered, if one made a soup of Turing machines and let them collide; one collision partner would act as the machine, and the other partner in the collision would act as the input tape. The soup of programs would act on itself, rewriting each other's programs, until . . . Until what?

Well, it didn't work. Many Turing machine programs are able to enter infinite loops and "hang." In such a case, the collision partners become locked in a mutual embrace that never ends, yielding no "product" programs. This attempt to create a silicon self-reproducing spaghetti of programs failed. Oh well.

On the wall at the Santa Fe Institute hangs a cartoon. It shows a rather puzzled fuzzy-headed kid pouring fluid into a beaker, a mess all over a table, and feathers flying all over the room. The caption reads: "God as a kid on his first try at creating chickens." Maybe before the Big Bang, the chief architect practiced on some other universe.

Fontana arrived at the institute, full of RNA landscapes, but like most inventive scientists, found a way to follow his more radical dream. If Turing machines "hung" when operating on one another, he would move to a similar mathematical structure called the lambda calculus. Many of you know one of the offspring of this calculus, for it is the programming language called Lisp. In Lisp, or the lambda calculus, a function is written as a string of symbols having the property that if it attempts to "operate" on another string of symbols, the attempt is almost always "legitimate" and does not "hang." That is, if one function operates on a second function, there is almost always a "product" function.

More simply, a function is a symbol string. Symbol strings operate on symbol strings to create new symbol strings! Lambda calculus and Lisp are generalizations of chemistry where strings of atoms—called molecules—act on strings of atoms to give new strings of atoms. Enzymes act on substrates to give products.

Naturally, since Fontana is a theoretical chemist, and since lambda and Lisp expressions carry out algorithms and since Fontana wanted to make a chemical soup of such algorithms, he called his invention algorithmic chemistry, or Alchemy.

I think Fontana's Alchemy may be an actual alchemy that begins to transform how we think of the biological, economic, and cultural world. You see, we can use symbol strings as models of interacting chemicals, as models of goods and services in an economy, perhaps even as models of the spread of cultural ideas—what the biologist Richard Dawkins called "memes." Later in the chapter, I will develop a model of technological evolution in which symbol strings stand for goods and services in an economy—hammers, nails, assembly lines, chairs, chisels, computers. Symbol strings, acting on one another to create symbol strings, will yield a model of the coevolution of technological webs where each good or service lives in niches afforded by other goods or services. In a larger context, symbol-string models may afford a novel and useful way to think about cultural evolution and the emergence of a global civilization, as we deploy ideas, ideals, roles, memes to act on one another in a never-ending unfolding rather like the molecular "wave front" expanding outward in a supracritical biosphere. Walter created our first mathematical language to explore the cascading implications of creation and annihilation that occurs in "soups" of these symbol strings as they act on and transform one another.

What happened when Fontana infected his computer with his Alchemical vision? He got collectively autocatalytic sets! He evolved "artificial life."

Here is what Walter did in his early numerical experiments. He created a "chemostat" on the computer that maintained a fixed total number of symbol strings. These strings bumped into one another just like chemicals. At random, one of the two strings was chosen as the program; the other, as the data. The symbol-string program acted on the symbol-string data to yield a new symbol string. If the total number of symbol strings exceeded some maximum, say 10,000, Fontana randomly picked one or a few and threw them away, maintaining the total at 10,000. By throwing away random symbol strings, he was supplying a selection pressure for symbol strings that were made often. By contrast, symbol strings that were rarely made would be lost from the chemostat.

When the system was initiated with a soup of randomly chosen symbol strings, at first these operated on one another to create a kaleidoscopic swirl of never-before-seen symbol strings. After a while, however, one began to see the creation of strings that had been encountered before. After a while, Fontana found that his soup had settled down to a self-maintaining ecology of symbol strings, an autocatalytic set.

Who would have expected that a self-maintaining ecology of symbol strings would come popping out of a bunch of Lisp expressions colliding and "rewriting" one another? From a random mixture of Lisp expressions, a self-maintaining ecology had self-organized. Out of nothing. What had Fontana found?

He had stumbled into two types of self-reproduction. In the first, some Lisp expression had evolved as a general "copier." It would copy itself, and copy anything else. Such a highly fit symbol string rapidly reproduced itself and some hangers-on and took over the soup. Fontana had evolved the logical analogue of an RNA polymerase itself made of RNA, hence a ribozyme RNA polymerase. Such an RNA would be able to copy any RNA molecule, including itself. Remember, Jack Szostak at Harvard Medical School is trying to evolve just such a ribozyme RNA polymerase. It would count as a kind of living molecule.

But Fontana found a second type of reproduction. If he "disallowed" general copiers, so they did not arise and take over the soup, he found that he evolved precisely what I might have hoped for: collectively autocatalytic sets of Lisp expressions. That is, he found that his soup evolved to contain a "core metabolism" of Lisp expressions, each of which was formed as the product of the actions of one or more other Lisp expressions in the soup.

Like collectively autocatalytic sets of RNA or protein molecules, Walter's collectively autocatalytic sets of Lisp expressions are examples of functional wholes. The holism and functionality are entirely nonmystical. In both cases, "catalytic closure" is attained. The whole is maintained by the action of the parts; but by its holistic catalytic closure, the whole is the condition of the maintenance of the parts. Kant would, presumably, have been pleased. No mystery here, but a clearly emergent level of organization is present.

Fontana called copiers "level-0 organizations" and autocatalytic sets "level-1 organizations." Now working with the Yale biologist Leo Buss, he hopes to develop a deep theory of functional organization and a clear notion of hierarchies of organizations. For example, Fontana and Buss have begun to ask what occurs when two level-1 organizations interact by exchanging symbol strings. They find that a kind of "glue" can be formed between the organizations such that the glue itself can help maintain either or both of the participating level-1 organizations. A kind of mutualism can emerge naturally. Advantages of trade and an economy are already implicit in coevolving level-1 autocatalytic sets.

Technological Coevolution

The car comes in and drives the horse out. When the horse goes, so does the smithy, the saddlery, the stable, the harness shop, buggies, and in your West, out goes the Pony Express. But once cars are around, it makes sense to expand the oil industry, build gas stations dotted over the countryside, and pave the roads. Once the roads are paved, people start driving all over creation, so motels make sense. What with the speed, traffic lights, traffic cops, traffic courts, and the quiet bribe to get off your parking ticket make their way into the economy and our behavior patterns.

The economist Joseph Schumpeter had spoken of gales of creative destruction and the role of the entrepreneur. But this was not the august Schumpeter speaking; it was my good friend, Irish economist Brian Arthur, hunched over seafood salad at a restaurant called Babe's in Santa Fe. Defying any theorem about rational choice, he always chose Babe's seafood salad, which, he said, tasted absolutely awful. "Bad restaurant," he vowed each time. "Why do you keep ordering their seafood salad?" I asked. No answer. It was the only time I had stumped him in seven years. Brian, among other things, has become deeply interested in the problem of "bounded rationality," why economic agents are not actually infinitely rational, as standard economic theory assumes, although all economists know the assumption is wrong. I suspect Brian is interested in this problem because of his own incapacity to try Babe's hamburgers, which are fine. Good restaurant.

"How do you economists think about such technological evolution?" I asked. From Brian and a host of other economists, I have begun to learn the answer. These attempts are fine and coherent. Work initiated by Sidney Winter and Dick Nelson and now carried on by many others center on ideas about investments leading to innovations and whether firms should invest in such innovations or should imitate others. One firm invests millions of dollars in innovation, climbing up the learning curve of a technological trajectory, as we discussed in Chapter 9. Other firms may also invest in innovation themselves, or simply copy the improved widget. IBM invested in innovation; Compaq cloned and sold IBM knock-offs. These theories of technological evolution are therefore concerned with learning curves, or rate of improvement of performance of a technology as a function of investment, and optimal allocation of resources between innovation and imitation among competing firms.

I am not an economist, of course, even though I have now enjoyed listening to a number of economists who visit the institute. But I cannot help feeling that the economists are not yet talking about the very facts that Brian Arthur first stressed to me. The current efforts ignore the fact that technological evolution is actually coevolution. Entry of the car drove the smithy to extinction and created the market for motels. You make your living in a "niche" afforded by what I and others do. The computer-systems engineer is making a living fixing widgets that did not exist 50 years ago. The computer shops that sell hardware make a living that was impossible to make 50 years ago.

Almost all of us make livings in ways that were not possible for *Homo habilis,* squatting around his fire, or even Cro-Magnon, crafting the magnificent paintings in Lascaux in the Périgord of southern France. In the old days, you hunted and gathered to get dinner each day. Now theoretical economists scratch obscure equations on whiteboards, not blackboards any longer, and someone *pays* them to do so! Funny way to catch dinner, I say.

(I was recently in the Périgord and purchased a flint blade made using upper Paleolithic techniques from a craftsman in Les Eyzies, near the Font-de-Gaume cave. In his mid-forties, with a horned callus half an inch thick on his right hand from hefting his hammer, an elk leg bone, he may be the singular member of our species who has made the largest number of flint artifacts in the past 60,000 years. But even he is making his living in a new niche—hammering flint for sale to the tourists awestruck by the Cro-Magnon habitat of our ancestors.)

We live in what might be called an economic web. Many of the goods and services in a modern economy are "intermediate goods and services," which are themselves used in the creation of still other goods and services that are ultimately utilized by some final consumers. The inputs to one intermediate good—say, the engine for a car—may come from a variety of other sources, from toolmakers to iron mines to computer-assisted engine design to both the manufacturer of that computer and the engineer who created the software to carry out computer-aided design. We live in a vast economic ecology where we trade our stuff to one another. A vast array of economic niches exist.

But what creates those niches? What governs the structure of the economic web, the arrangement by which jobs, tasks, functions, or products connect with other jobs, tasks, functions, or products to form the web of production and consumption?

And if there is an economic web, surely it is more complex and tangled now than during the upper Paleolithic when Cro-Magnon painted. Surely it is more complex now than when Jericho first built its walls. Surely it is more complex than when the Anasazi of New Mexico created the Chacoan culture 1,000 years ago. If the economic web grows more tangled and complex, what governs the structure of that web? And the question I find most fascinating: If the economy is a web, as it surely is, does the structure of that web itself determine and drive how the web transforms? If so, then we should seek a theory of the self-transformation of an economic web over time creating an ever-changing web of production technologies. New technologies enter (like the car), drive others extinct (like the horse, buggy, and saddlery), and yet create the niches that invite still further new technologies (paved roads, motels, and traffic lights).

The ever-transforming economy begins to sound like the ever-transforming biosphere, with trilobites dominating for a long, long run on Main Street Earth, replaced by other arthropods, then others again. If the patterns of the Cambrian explosion, filling in the higher taxa from the top down, bespeak the same patterns in early stages of a technological trajectory, when many strong variants of an innovation are tried until a few dominant designs are chosen and the others go extinct, might it also be the case that the panorama of species evolution and coevolution, ever transforming, has its mirror in technological coevolution as well? Maybe principles deeper than DNA and gearboxes underlie biological and technological coevolution, principles about the kinds of complex things that can be assembled by a search process, and principles about the autocatalytic creation of niches that invite the innovations, which in turn create yet further niches. It would not be exceedingly strange were such general principles to exist. Organismic evolution and coevolution and technological evolution and coevolution are rather similar processes of niche creation and combinatorial optimization. While the nuts-and-bolts mechanisms underlying biological and technological evolution are obviously different, the tasks and resultant macroscopic features may be deeply similar.

But how are we to think about the coevolving web structure of an economy? Economists know that such a structure exists. It is no mystery. One did not have to be a financial genius to see that gas stations were a great idea once cars began to crowd the road. One hustled off to one's friendly banker, provided a market survey, borrowed a few grand, and opened a station.

The difficulty derives from the fact that economists have no obvious way to build a theory that incorporates what they call complementarities. The automobile and gasoline are consumption complementarities. You need both the car and the gas to go anywhere. When you tell the waitress, "Make it ham and eggs, over easy," you are specifying that you like ham with your eggs. The two are consumption complements. If you went out with your hammer to fasten two boards together, you would probably pick up a few nails along the way; hammer and nails are production complements. You need to use the two together to nail boards together. If you chose a screwdriver, you would feel rather dumb picking up some nails on your way to your shop to make a cabinet. We all know screws and screwdrivers go together as production complements. But nail and screw are what economists call production substitutes. You can usually replace a nail with a screw and vice versa.

The economic web is precisely defined by just these production and consumption complements and substitutes. It is just these patterns that create the niches of an economic web, but the economists have no obvious way to build a "theory" about them. What would it mean to have a theory that hammer and nail go together, while car and gasoline go together? What in the world would it mean to have a theory of the functional connections between goods and services in an economic web? It would appear we would have to have a theory of the functional couplings of all possible kinds of widgets, from windshield wipers to insurance policies to "tranches" of ownership in bundled mortgages, to laser usage in retinal surgery. If we knew the "laws" governing which goods and services were complements and substitutes for one another, we could predict which niches would emerge as new goods were created.

We could build a theory about how the technological web drives its own transformation by persistent creation of new niches.

Here is a new approach. What if we think of goods and services as symbol strings that we humans can use as "tools," "raw materials," and "products?" Symbol strings act on symbol strings to create symbol strings. Hammer acts on nail and two boards to make attached boards. A protein enzyme acts on two organic molecules to make two attached molecules. A symbol string, thought of in Lisp light, is both a "tool" and a "raw material" that can be acted on by itself or other tools to create a "product." Hence the Lisp laws of chemistry implicitly define what are production or consumption complements or substitutes. Both the "enzyme" and the "raw-material" symbol strings are *complements* used in the creation of the product symbol string. If you can find another symbol-string "enzyme" that acts on the "raw-material" string to yield the same product, then the two enzyme strings are *substitutes.* If you can find a different raw-material string that, when acted on by the original enzyme string, yields the same final product, then the two raw-material strings are substitutes. If the outputs of one such operation yield products that enter into other production processes, you have a model of a webbed set of production functions with complementarities and substitutes defined implicitly by Lisp logic. You have the start of a model of functionally coupled entities acting on one another and creating one another. You have, in short, the start of a model of an economic web where the web structure drives its own transformation.

The web of technologies expands because novel goods create niches for still further new goods. Our symbol-string models therefore become models of niche creation itself. The molecular explosion of supracritical chemical systems, the Cambrian explosion, the exploding diversity of artifacts around us today—all these drives toward increased diversity and complexity are underpinned by the ways each of these processes creates niches for yet further entities. The increase in diversity and complexity of molecules, living forms, economic activities, cultural forms—all demand an understanding of the fundamental laws governing the autocatalytic creation of niches.

If we do not have the real laws of economic complementarity and substitutability, why hammers go with nails and cars with gasoline, what's the use of such abstract models? The use, I claim, is that we can discover the kinds of things that we would expect in the real world if our "as if" mock-up of the true laws lies in the same universality class. Physicists roll out this term, "universality class," to refer to a class of models all of which exhibit the same robust behavior. So the behavior in question does not depend on the details of the model. Thus a variety of somewhat incorrect models of the real world may still succeed in telling us how the real world works, as long as the real world and the models lie in the same universality class.

Random Grammars

At about the same time as Alonzo Church developed the lambda calculus as a system to carry out universal computation and Alan Turing developed the Turing machine for the same purpose, another logician, Emil Post, developed yet another representation of a system capable of universal computation. All such systems are known to be equivalent. The Post system is useful as one approach to trying to find universality classes for model economies.

On the left and right sides of the line in Figure 12.1 are pairs of symbol strings. For example, the first pair has (111) on the left and (00101) on the right. The second pair has (0010) on the left and (110) on the right. The idea is that this list of symbol strings constitutes a "grammar." Each pair of symbol strings specifies a "substitution" that is to be carried out. Wherever the left symbol string occurs, one is to substitute the right symbol string. In Figure 12.2, I show a "pot" of symbol strings on which the grammar in Figure 12.1 is supposed to "act." In the simplest interpretation, you apply the grammar of Figure 12.1 as follows. You randomly pick a symbol string from the pot. Then you randomly choose a pair of symbol strings from the figure. You try to match the left symbol string in the figure with the symbol string you chose. Thus if you picked the first pair of symbol strings in the figure and you find a (111) in the symbol string you chose from the pot, you are to "snip" it out and substitute the right symbol string from the same row of Figure 12.1. Thus (111) is replaced by (00101).

Obviously, you might continue to apply the grammar rules of Figure 12.1 to the symbol strings in the pot ad infinitum. You might continue

Grammar Table

1	1	1	0 0 1 0 1			
0 0 1 0			1 1 0			
0 0			1 0 1 1			
1 0 0 1			0 1			
1 0 1			0 0 1 0			

Figure 12.1 *A Post grammar. Instances of the left symbol string are to be replaced by the corresponding right symbol string.*

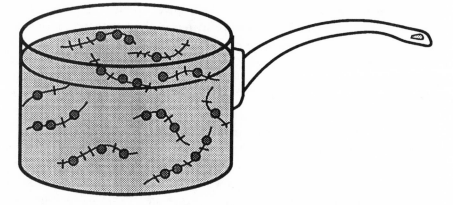

Figure 12.2 *When the Post grammar in Figure 12.1 is applied to a "pot" of symbol strings, a succession of new symbol strings emerges.*

to randomly pick up symbol strings in the pot, choose a row from the figure, and apply the corresponding substitution. Alternatively, you might define precedence rules about the order in which to apply the rows to any symbol string. And you might notice that sometimes the application of a substitution in Figure 12.1 to a symbol string in Figure 12.2 would create a new "site" in the symbol string, which was itself a candidate for application of the rule that just created it. To avoid an infinite loop, you might decide to apply any row substitution from Figure 12.1 to any site only once before all the other rows had been chosen.

Any such set of decisions about application of the substitutions in Figure 12.1 plus decisions about the order of applying the rules to the symbol strings yields a kind of algorithm, or program. Starting with a set of initial symbol strings in the pot, you would keep applying the substitutions and derive a sequence of symbol strings. Like a Turing machine converting an input tape to an output tape, you would have carried out some kind of computation.

Now the next step is to remove you, the reader, from the action, and allow the symbol strings in the pot to act on one another, like enzymes do on substrates, to carry out the substitutions "mandated" by the "laws of substitution" in Figure 12.1. An easy way to do this is to define "enzymatic sites." For example, the first row of Figure 12.1 shows that (111) is to be transformed to (00101). Let us think of a symbol string in the pot of Figure 12.2 with a (111) sequence somewhere in it as a substrate. An "enzyme" might be a symbol string in the same pot with a "template matching" (000) site somewhere in it. Here the "enzyme match rule" is that a 0 on the enzyme matches a 1 on the substrate, rather like nucleotide base-pairing. Then given such a rule for "enzymatic sites," we can allow the symbol strings in the pot to act on one another. One way is to imagine two randomly chosen symbol strings colliding. If either string has an "enzymatic site" that matches a "substrate site" on the other, then the enzymatic site "acts on" the substrate site and carries out the substitution mandated in the corresponding row of Figure 12.1.

That's all there is to it. Now we have an algorithmic chemistry specified by a specific "grammar." The symbol strings in the pot transform one another into new symbol strings, which again act on one another, ad infinitum. This persistent action will generate some flowering of symbol strings. It is the behavior of this florescence of symbol strings over time that is of interest now. For the patterns of flowering are to become our models of technological coevolution. To accomplish this, we will have to add a few ideas.

But first, how should we pick our grammar, as exemplified by Figure 12.1? No one knows "the right way" to specify the choice of pairs of symbol strings in the "laws of substitution" figure. Worse, there is an infinity of possible choices! In principle, the number of pairs of symbol strings might be infinitely long. Moreover, no one is limiting us to single symbol strings as "enzymes" acting on single symbol strings as "substrates" to yield single symbol strings as "products." We can perfectly well think about an ordered set of symbol strings as an "input bundle" and an ordered set of symbol strings as a "machine." Push the input bundle into the machine, and you get some "output bundle." The "machine" would be like an assembly line, doing a series of transformations on each input symbol string.

If we want to allow input bundles and machines, and if each can be any subset of the symbol strings, then a mathematical theorem says that the number of possible grammars is not just infinite, but is second-order infinite. That is, the number of possible grammars, like the real numbers, is uncountably infinite.

Since we cannot count the possible grammars, let's cheat. Let's just sample grammars at random from the infinity of possible grammars and find out what grammars in different regions of "grammar space" do. Let's imagine that we can find regions of grammar space within which the resulting behavior of our pot of symbol strings is insensitive to the details. Let's look, in short, for universality classes.

One way to define classes, or ensembles, of grammars in regions of grammar space is by the number of pairs of symbol strings that can occur in the grammar, the distributions of their lengths, and the way the longer and shorter symbol strings are distributed as the left or right member of the pair. For example, if all the right members are smaller than the left members, substitution will eventually lead to very short symbol strings, which are too short to match any "enzymatic site." The "soup" will become inert. In addition, the complexity of allowed "input bundles," "machines," and "output bundles" can be defined in terms of the number of symbol strings in each. As these parameters defining grammars are systematically altered, different regions of grammar space can be explored. Presumably, different regions will give rise to different characteristic behaviors. These different regions and behaviors would be the hoped-for universality classes.

This systematic study has not yet been done. If we could find a region of "grammar space" that gave us models of technological coevolution that look like real technological coevolution, then perhaps we might have found the right universality class and the correct "as if" model of the unknown laws of technological complementarity and substitutability. Here, then, is a program of research.

The program has begun, for my colleagues and I have actually made a few small models of economies that are already yielding interesting results.

Eggs, Jets, and Mushrooms

Before we turn to economic models, let us consider some of the kinds of things that can happen in our pot of symbol strings as they act on one another, according to the laws of substitution we might happen to choose. A new world of possibilities lights up and may afford us clues about technological and other forms of evolution. Bear in mind that we can consider our strings as models of molecules, models of goods and services in an economy, perhaps even models of cultural memes such as fashions, roles, and ideas. Bear in mind that grammar models give us, for the first time, kinds of general "mathematical" or formal theories in which to study what sorts of patterns emerge when "entities" can be both the "object" acted on and transformed and the entities that do the acting, creating niches for one another in their unfolding. Grammar models, therefore, help make evident patterns we know about intuitively but cannot talk about very precisely.

One might get a symbol string that copied itself or any other symbol string, a kind of replicase.

One might get a collectively autocatalytic set of symbol strings. Such a set would make itself from itself. In *The Origins of Order*, I used a name I thought of late one night. Such a closed self-creating set is a kind of "egg" hanging in the space of symbol strings.

Suppose that one had a perpetually sustained "founder set" of symbol strings. Such a sustained founder set of symbol strings might create new symbol strings, which in turn acted on one another to create still new strings—say, ever longer symbol strings—in a kind of "jet." A jet would squirt away from the founder set out into the space of possible strings.

The jet might be finite or infinite. In the latter case, the founder set would squirt out a jet of ever-increasing diversity of symbol strings.

An egg might be leaky and squirt out a jet. Such an egg object would hang like some bizarre spaceship, spraying a jet of symbol strings into the inky blackness of the far reaches of string space.

A sustained founder set might create a jet whose symbol strings were able to "feed back" to create symbol strings initially formed earlier by alternative routes. In my late-night amusement, I called these "mushrooms." A mushroom is a kind of model for "bootstrapping." A symbol string made by one route can later be made by another route via a second symbol string that the first one may have helped create. Stone hammers and digging tools became refined, eventually led to mining and metallurgy, which led to the creation of machine tools, which now manufacture the metal tools used to make the machine tools. Hmm. Bootstrapping. Think, then, about how common mushrooms must be in our technological evolution since the lower Paleolithic. The tools we make help us make tools that in turn afford us new ways to make the tools we began with. The system is autocatalytic. Organisms and their collectively autocatalytic metabolisms built on a sustained founder set of exogenously supplied food and energy are kinds of mushrooms, as is our technological society. Mushroom webs in ecosystems and economic systems are internally coherent and "whole." The entities and functional roles that each plays meet and match one another systematically.

A sustained founder set of symbol strings might make an infinite set of symbol strings, but there may be a certain class of symbol strings that will never be made from that founder set. For example, it may be the case that no symbol string starting with the symbols (110101 . . .) will ever be made. While an infinite set is made, an infinite number are never made. Worse, given the initial set of symbol strings and the grammar, it can be formally impossible to prove or disprove that a given symbol string, say (1101010001010), will never be produced. This is called *formal undecidability* in the theory of computation and is captured in Kurt Gödel's famous incompleteness theorem.

In a moment, we're going to imagine that we live in such a world. Formal undecidability means that we cannot, in principle, predict certain things about the future. Perhaps we cannot predict, for example, if we lived in the world in question, that (11010100001010) will never be formed. What if (11010100001010) were Armageddon? You'd never know.

What if it is true that the technological, economic, and cultural worlds we create are genuinely like the novel string worlds we are envisioning? After all, string worlds are built on the analogy with the laws of chemistry. If one could capture the laws of chemistry as a formal grammar, the stunning implication of undecidability is this: given a sustained founder set of chemicals, it might be formally undecidable whether a given chemical could not possibly be synthesized from the founder set! In short, the underlying laws of chemistry are not mysterious, if not entirely known. But if they can be captured as a formal grammar, it is not unlikely, and indeed is strongly urged by Gödel's theorem, that there would remain statements about the evolution of a chemical reaction system that would remain impossible to prove or disprove.

Now, if formal undecidability can arise from the real laws of chemistry, might the same undecidability not arise in technological or even cultural evolution? Either we can capture the unknown laws of technological complementarity and substitutability in some kind of formal grammar or we cannot. If we can, then Gödel's theorem suggests that there will be statements about how such a world evolves that are formally undecidable. And if we cannot, if there are no laws governing the transformations, then surely we cannot predict.

Technological Coevolution and Economic Takeoff

In helping to build theories of technological coevolution, I suspect that these toy worlds of symbol strings may also reveal a new feature of technological evolution: subcritical and supracritical behavior. A critical diversity of goods and services may be necessary for the sustained explosion of further technological diversity. Standard macroeconomic models of growth are single-sector models—a single stuff is produced and consumed. Growth is seen as increasing amounts of stuff produced and consumed. Such a theory envisions no role for the growth of diversity of goods and services. If this growth in diversity is associated with economic growth itself, as is supported by some evidence, then diversity itself may bear on economic take off.

In Figure 12.3, I show an outline of France. Again, we are going to model goods and services as symbol strings. We want to think about how the "technological frontier" evolves as the French people realize the unfolding potential of the raw materials with which we are about to

Figure 12.3 *The outline of France, showing different products growing from the soil. These symbol strings represent the renewable resources—wood, coal, wool, dairy, iron, wheat, and such—with which France is endowed. As the people use the strings to act on other strings, new, more complex products emerge.*

endow them. Imagine that, each year, certain kinds of symbol strings keep emerging out of the fertile soil of France. These symbol strings are the "renewable resources" of France, and might stand for grapes, wheat, coal, milk, iron, wood, wool, and so forth. Now let's forget the values of any of these goods and services, the people who will work with them, and the prices that must emerge and so forth. Let's just think about the evolving "technological possibilities" open to France, ignoring whether anyone actually wants any of the goods and services that might be technically feasible.

At the first period, the French might consume all their renewable resources. Or they might consult the "laws of technological complementarity" engraved at the *hôtel de ville* in each town and village, and consider all the possible new goods and services that might be created by using the renewable resources to "act" on one another. The iron might be made into forks, knives, and spoons, as well as axes. The milk might be made into ice cream. The wheat and milk might be made into a porridge. Now at the next period, the French might consume what they had by way of renewable resources, and the bounty of their first inventions, or they might think about what else they could create. Perhaps the ice cream and the grapes can be mixed, or the ice cream and grapes mixed and placed in a baked shell made of wheat to create the first French pastry. Perhaps the axe can be used as such to cut firewood. Perhaps the wood and axe can be used to create bridges across streams.

You get the idea. At each period, the goods and services previously "invented" create novel opportunities to create still more goods and services. The technological frontier expands. It builds on itself. It unfolds. Our simple grammar models supply a way to talk about such unfolding.

Economists like to think about at least slightly more complex models that include consumers and their demands for the potential goods and services. Imagine that each symbol string has some value, or utility, to the one and only consumer in French society. The consumer might be Louis XIV, or Jacques the good hotelier, or actually a number of identical French folks with the same desires. In this simple model, there is no money and there are no markets. In their place is an imaginary wise social planner. The task of the social planner is this: she knows the desires of Louis XIV (or she had better), she knows the renewable resources of the kingdom, and she knows the "grammar table," so she can figure out what the ever-evolving technological frontier will look like by "running the projection" forward. All she has to do is to try to optimize the overall happiness of the king, or all the Jacques, over the future. At this point, simple economic models posit something like this: the king would rather have his pleasures today, thank you, than at any time in the future. Indeed, if he has to wait a year, he is willing to trade X amount of happiness now for X plus 6 percent next year, and 6 percent more for each year he must wait. In short, the king discounts the value of future utility at some rate. So does Jacques. So do you.

So our infinitely wise social planner thinks ahead some number of periods, say 10, called a planning horizon; thinks through all the possible sequences of technological goods and services that might possibly be created over these 10 periods; considers the (discounted) happiness of the king for all these possible worlds; and picks a plan that makes his lordship as pleased as possible. This plan specifies how much of each technologically possible "production" is actually to be carried out over the 10 periods and how much of what will be consumed when. These production activities occur in some ratio: 20 times as much ice cream and grapes as axes, and such. The ratio actually is the analogue of price, taking one of the goods as "money." Not all of the possible goods may be produced. They do not make the king happy enough to waste resources on. Thus once we include the utilities of the goods and services, the goods actually produced at any moment will be a subset of the entire set of what is technologically possible.

Now the social planner implements the first year of her plan, the economy cranks forward for a year with the planned production and consumption events, and she makes a new 10-year plan, now extending from year 2 to year 11. We've just considered a "rolling horizon" social planner model. Over time, the model economy evolves its way into the future. At each period, the social planner plans ahead 10 periods, picks the optimal 10-year plan, and implements the first year.

Such models are vast oversimplifications, of course. But you can begin to intuit what such a model in our grammar context will reveal. Over time, novel goods and services will be invented, displacing old goods and services. Technological speciation and extinction events occur. Because the web of technologies is connected, the extinction of one good or service can start a spreading avalanche in which other goods and services no longer make sense and lapse from view. Each represented a way of making a living, come and gone, having strutted its hour. The set of technologies unfolds. The goods and services in the economy not only evolve, but coevolve, for those present must always make sense in the context of the others that already exist.

Thus these grammar models afford us new tools to study technological coevolution. In particular, once one sees one of these models, the idea that the web itself drives the ways the web transforms becomes obvious. We know this intuitively, I think. We just have never had a toy world that exhibited to us what we have always known. Once one sees one of these models, once it becomes obvious that the economic web we live in largely governs its own directions of transformations, one begins to think it would be very important indeed to understand these patterns in the real economic world out there.

These grammar models also suggest a possible new factor in economic takeoff: *diversity probably begets diversity; hence diversity may help beget growth.*

In Figure 12.4, I show, on the x-axis, the diversity of goods and services that emerge from the French soil as renewable resources each period. On the y-axis, I show the number of pairs of symbol strings com-

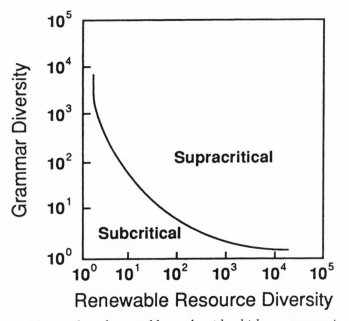

Figure 12.4 *The number of renewable goods with which an economy is endowed is plotted against the number of pairs of symbol strings in the grammar, which captures the hypothetical "laws of substitutability and complementarity." A curve separates a subcritical regime below the curve and a supracritical regime above the curve. As the diversity of renewable resources or the complexity of the grammar rules increases, the system explodes with a diversity of products.*

posing the grammar, or laws, of complementarity and substitutability. And in this xy plane, I sketch the now familiar curve separating the now familiar subcritical and supracritical behaviors.

Imagine that the grammar laws had only a single pair of symbol strings. Imagine that an impoverished France sprouted only a single kind of symbol string from its soil each spring. Alas, the grammar law might be such that nothing at all could be done with the single renewable resource to make anything new and interesting. All the French could do is to consume that resource. No explosion of the technological frontier could occur. If by dint of hard work, the French saved excess amounts of this single resource, well that's good. Nevertheless, no explosion of diverse goods could occur. The system is subcritical.

But suppose the grammar laws have many pairs of symbol strings, and the fertile soil of France sprouts many kinds of renewable resources. Then the chances are overwhelming that a large number of useful and interesting products can be created immediately from these founder symbol strings by using them to transform one another. In turn, the enhanced diversity of goods and services can lead to a further explosion of the technological frontier. If the social planner deems them useful to the king, an unfolding subset of the technically possible goods and services will actually arise and extinguish in some complex progression. Economic takeoff in diversity has occurred. The system is supracritical.

If France were subcritical and so, too, were England across the Channel, then trade between them might be sufficient to allow the two to become technologically supracritical. So a larger, more complex economy may grow in diversity because it allows the technological frontier to explode.

The behavior of our model economies also depends on the "discount" factor and the social planner's planning horizon. If the king does not want to wait for his happiness, then the wise social planner does not put off drinking milk today. Ice cream is never created. The economy that might have blossomed into diversity remains truncated, perhaps blissfully, in its initial Garden of Eden state. Alternatively, whatever the king might prefer, if the social planner does not think ahead, she never realizes that ice cream can be created. Again, the model economy truncates its diversification.

Models with social planners, while used by economists, are far less realistic than models in which there are markets and economic agents who buy and sell. In social-planner models, all the problems of coordination of actions among the agents, the invisible hand part of economics, is taken care of by the planner, who simply commands the appropriate ratio of the different production and consumption activities. In the real world, independent agents make decisions and the market is supposed to coordinate their behaviors. While the social-planner model I have used ignores all the important issues concerning the emergence of markets and behavior coordination among such agents, I have done so to concentrate on the evolution of the web of technologies. Birth and extinction of technologies, and subcritical and supracritical behaviors can occur. It seems reasonable, but remains to be shown, that similar features will show up in more realistic versions of this kind of model in which the social planner is replaced with markets and optimizing agents.

Caveats. I am not an economist. These grammar models are new. You should at most take them as metaphors at present. Yet even as metaphors, they make suggestions that may be worth investigation. Among these, the possibility that diversity may help drive economic growth.

Standard theories of economic growth appear not to have taken into account the potential linkages between the diversity of economic sectors in economic growth. Standard macroeconomic theories often build models of economic growth based on an economy producing a single stuff, a kind of aggregate of all our productions, and talk in terms of aggregate demand, aggregate supply, money growth, interest rate, and other aggregated factors. Long-term economic growth is typically attributed to two major factors: technological improvements and growth in the productivity and skill of workers, called human capital. Growth in technology is seen as occurring in response to investments in research and development. Growth in human capital occurs because of investments in education and on-the-job learning. Here the improvements accrue to the benefit of the individual or his immediate family. How "technological improvements" and "human capital" may be linked to the underlying dynamics of technological webs and their transformation is not yet well articulated.

It is not that economists are unaware of the kinds of complementarities we have discussed. Indeed, enormous input–output matrices of economic interaction are studied. But lacking a formalizable framework, economists appear to have had no obvious way to build models of connections between various economic sectors and study their implications for further diversification and economic growth. Yet there is beginning to be evidence of the importance of these cross-connections. If this view is correct, then diversity should be a major predictor of economic growth. This is not a new idea. Canadian economist Jane Jacobs advanced the same idea on different grounds two decades ago. Recently, University of Chicago economist José Schenkman, also a Santa Fe Institute friend, reported work that strongly suggests that the rate of economic growth of cities is, in fact, strongly correlated with the diversity of the sectors within the cities. Schenkman and his colleagues carefully controlled for the aggregate capitalization of the industries and the specific sectors involved. Thus at least some clues support the rather obvious idea we discuss here: the web structure of an economic system is itself an essential ingredient in how that economic system grows and transforms.

Indeed, if we tie together the potential implications of the parallels between the coevolution of organisms and of artifacts that we have discussed in the previous chapters, something of a new, perhaps useful, framework concerning aspects of economic growth begins to emerge.

We have already seen that it is characteristic of optimization of conflict-laden problems that improvement is rapid initially, and then slows exponentially. This well-known feature of technological learning curves implies that after a major innovation, there can be an early period of increasing returns. A given investment in the technology increases productivity greatly. Later, as improvement slows exponentially, further investment faces diminishing returns. This suggests that capital and credit will flow into the new sector in the early stages, during the increasing-returns phase. If so, then major innovations drive capital formation and growth in the sector they create. Just this is occurring in biotechnology today. Later, as the learning curve is climbed and as markets saturate, growth in the mature sector dwindles.

But as the economic activities alter, the coevolutionary economic landscape deforms. A family of new, "neighboring" technologies will proliferate, climbing uphill on the deformed landscapes. As aircraft design and engine power increased, the fixed-blade propeller became less useful than a new innovation, the variable-pitch propeller. Like branching speciation on a deforming landscape where the newly formed species may initially be relatively unfit, the newly invented variable-pitch propeller opened a modest new era of learning how to create better variable-pitch propellers. So as nearby products and technologies are called forth by the deforming landscapes, each generates a new burst of rapid learning, and a burst of increasing returns can attract capital and credit, driving further growth in that sector. In addition, the learning-by-doing that occurred with fixed-pitch propellers spills over to the neighboring new technologies. The human capital, the learned skills, are naturally accumulated on a wider basis than the individual and his family, or even than a single narrow technology.

On a larger scale, persistent innovation in an economy may depend fundamentally on its supracritical character. New goods and services create niches that call forth the innovations of further new goods and services. Each may unleash growth both because of increasing returns in the early phase of improvement on learning curves or new open markets. Some of these are truly triggers of Schumpeterian "gales of creative destruction," ushering out many old technologies, ushering in many new ones in vast avalanches. Such avalanches create enormous arenas of increasing returns because of the massive early improvements climbing learning curves along the novel technological trajectories, as well as major new markets. So such large avalanches drive significant capital formation and growth. Other new technologies come and go with hardly a ripple. These differences presumably reflect, in part, how central or peripheral the new technology or product is in the current web and in its future evolution. The automobile and the computer were central. The hula hoop was peripheral.

We have only sketched some obvious ideas, but they may be worth serious investigation. Diversity begets diversity, driving the growth of complexity. Such ideas might eventually have policy implications. If diversity matters, then helping Third World countries might be better accomplished by fostering cottage industries that create a local web that is mutually reinforcing and can take root and grow, rather than by creating the Aswan Dam. But such conclusions lie in the future. At best, I hope you will find grammar models an intriguing way to think about technological evolution and its roles in economic growth.

Before concluding our discussion of economic webs, I ask you to imagine yourself living in the world just sketched. If such systems forever unfold with technological innovations, then what is wise? If the side products of the technologies have unforeseeable long-term consequences for the planet, what is wisdom? Bell South needs to decide whether to invest billions in fiber-optic technologies. Should Bell South do so? What if, in two years, some bright kid thinks up a way to toss tin cans into the sky held up by fans placed at strategic intervals in such a way that fiber-optics is less useful. Billions down the tube. Can Bell South management be sure of what to do, given the unfolding technological frontier?

No.

Who would have dreamed, a decade ago, that today we would be faxing from home to home? Recently, I was invited to an intriguing meeting in Colorado, a few hundred miles from my home in Santa Fe. I had lost the detailed instructions, so called a friend, Joan Halifax, who was going. Joan was not at home. I left word on her answering machine. Ten minutes later, she called from the northern reaches of Vancouver Island, where she was hoping to spy a few whales. I explained my confusion. Within minutes, a fax arrived from Vancouver Island with instructions. Good grief.

Global positioning systems (GPS), now allow you, for a few hundred dollars, to buy a widget that locks into a few satellites and locates you on the surface of the earth within a few dozen feet. It is possible to locate to within a few inches or less, but the U.S. military apparently keeps the precision lower for civilian uses. Today I heard a rumor that the Japanese, properly concerned with earthquakes, are setting up such systems at fixed points on their islands. Small changes in the distances between these points would foretell changes that might themselves foretell an earthquake. I don't know if the rumor is true, but it is plausible. So we measure location on the earth and changes in location to guess the behavior of magma down there, by tossing signals up to satellites we tossed up before the signals. And Columbus thought himself lucky to have a lodestone. The web does unfold.

It is not merely that we must imagine what it would be like to live in our model unfolding grammar world. We do live in such a world. We live on a self-organized sandpile that sheds avalanches down the critical slopes with each footstep. We have hardly a clue what will unfold.

Global Civilization

What is to become of our patchwork of civilizations, ancient and new, drawn ever more tightly into one another's embrace? Like it or not, some form of global civilization will emerge. We are at that particular time in history when population, technology, economics, and knowledge spin us together. I write with no particular wisdom, but I, like you, am a member of this emerging civilization. I wonder if we really understand very much of what we are creating, of the plausible foundations such a way of being must have to hold us all with some modicum of tolerance and forgiveness of one another. I wonder if the political structures we have created will continue to serve us.

When Western culture touched the Inuit culture, the latter was soon vastly transformed. When Western culture touched traditional Japanese culture, the latter was soon vastly transformed. When Rome touched Athens, Rome was transformed and so was Athens. When the Hellenistic and Hebraic worlds collided, the cornerstones of Western civilization fell together in some new way. When the Spaniards touched the Aztecs, a new cultural mixture was formed in the crucible. Ferocious gods glower from frescos painted with Spanish skills. Guatemalan patterns linger in tapestries woven in Western styles.

Now all our cultures are in collision. At the small meeting in Nambe, New Mexico, where I met N. Scott Momaday, Lee Cullum, and Walter Shapiro, where we were asked by an organization called the Gihon Foundation to be presumptuous enough to think about the major problems confronting the world, my own overwhelming concerns focused on this emerging world civilization, the cultural dislocations it would engender, and the issue of whether we could find a cultural and political framework that would work for us.

Somehow, the string images we have discussed press themselves on me. The swirl of transformations of ideologies, fashions begetting fashions begetting fashions, cuisines begetting novel cuisines, legal codes and precedents begetting the further creation of law, seem similar in as yet unclear ways to model grammar worlds with their eggs, jets, and mushrooms. If a new symbol string is tossed casually into Fontana's silicon pot, a swarm of novel symbol strings might form. The small perturbation can yield a vast transformation in the future of the string system—or nothing at all.

When Mikhail Gorbachev began speaking of *glasnost,* we knew that something big might happen. We knew that a move to open the closed Soviet society to its own people's concerns might unlock a revolution. We knew that the small steps might lead to vast transformations. Yet while we knew this intuitively, and the pundits pummeled us with their insights, we do not really know what we intuited. We do not understand how the logs of the logjam fit together, such that removing one particular log will cause the logjam to shift only slightly, while another, innocuous in its position, will unleash the whole mass to swarm downriver. We do not understand the functional couplings among the elements—political, economic, cultural—of our public world.

When the Chinese government tragically decided to kill the young students on Tiananmen Square, those leaders feared the particular log that was being tugged by the students. Yet none of us really have much insight into the structure of that logjam either.

We lack a theory of how the elements of our public lives link into webs of elements that act on one another and transform one another. We call these transformations "history." Hence with all the accidents of history, biological and human, one must engage in a renewed debate: Is there a place for law in the historical sciences? Can we find lawlike patterns—cultural, economic, and otherwise—such as subcritical and supracritical behavior, or patterns of speciation and extinction?

We had best attempt to understand such processes, for the global civilization is fast upon us. We will live through its birth, ready or not.

Modern democracy as we encapsulate it, as we tell ourselves the story of ourselves, is so much a product of the Enlightenment. Newton and Locke. The United States Constitution, which has served so well for more than 200 years, is a document built on an image of balancing political forces holding an equilibrium. Newton and Locke. Our political system is built to be flexible enough to balance political forces and allow the polity to evolve. But our theory of democracy takes little account of the unfolding, evolving nature of cultures, economies, and societies. In the nineteenth century, the idea of historical science came to the fore. Hegel gave us thesis, antithesis, and synthesis. Marx stood Hegel on his idealist head to create dialectical materialism. These ideas now stand discredited. Yet thesis, antithesis, synthesis sounds more than a little bit like the evolution of the hundreds of millions of species that have come and gone, or the evolution of technologies that have come and gone.

John Maynard Smith, at a meeting at the Santa Fe Institute in the summer of 1992, startled me by telling a group gathered together, "Perhaps you all are embarking on some kind of post-Marxist analysis of social evolution." I did not know what he meant. Marxism has such a bad reputation that I wasn't certain I was at all happy with whatever it was he might have meant. Yet might we be beginning to develop the conceptual tools that may help us to understand a bit more of the historical evolution of societies? Historians do not think of themselves as merely recounting the record, but as looking for the web of influences that kaleidoscopically interact with one another to create the patterns that unfold. Is there, in fact, a place for "law" in the historical sciences? Is the Industrial Revolution, with its explosion of technologies, an example of assembling a critical diversity of goods and services, of new production technologies, such that the diversity fed on itself autocatalytically? What of cultural revolutions such as the Renaissance and the Enlightenment? Do these somehow reflect a web of collectively self-reaffirming ideas, norms, and agencies?

Richard Dawkins, an evolutionary biologist at Oxford, popularized the word *meme*. In the simplest image, a meme is a bit of behavior that is replicated in a population. Women now wear sunglasses perched on top of their heads. I suspect that this was spawned by Audrey Hepburn playing Holly Golightly in the movie *Breakfast at Tiffany's* many years ago. In this limited sense, memes are "replicators," which are invented and then imitated in complex patterns of cultural transmission. But this sense of meme is too limited. It is like one of Fontana's level-0 organizations, a mere replicator that may spread in a population. Are there collectively autocatalytic sets, level-1 organizations of memes? Can cultural patterns be thought of as self-sustaining and mutually defining sets of beliefs, behaviors, roles?

Perhaps at present, the analogy is loose, more of a metaphor than the start of a real theory. But might we not be members of a variety of such cultural entities? We invent concepts and categories that we use to carve up the world. Those categories are mutually defined in a complex reaffirming circle. How could it be otherwise? Having invented the categories, we carve the world into them and find ourselves categorized as well. Due to the creation of a legal system, I am able to enter into contracts. Because we can both do so, you and I can create a person that may live forever, the corporation, which takes on aims that survive and can even harm the interests of many of those who founded it. Thus the modern corporation is a collectively self-sustaining structure of roles and obligations that "lives" in an economic world, exchanges signals and stuffs, and survives or dies in ways at least loosely analogous to those of *E. coli*. *E. coli* is collectively autocatalytic and sustains itself in its world. The modern corporation also seems collectively autocatalytic. Both *E. coli* and IBM coevolve in their respective worlds. Both participate in the creation of the ecosystem each inhabits. As the recent difficulties of massive IBM show, even the mighty can find their world vastly transformed.

So a global civilization is emerging. At the small meeting in Nambe, Walter Shapiro (a political writer for *Time*) and Lee Cullum (a writer for the *Dallas Morning News*) focused on the implications of the advent of a McDonald's in the corner of an ancient castle in Heidelberg, a German university town. Good business? Probably. Unsettling, certainly. Maria Verela is a MacArthur Fellow living 50 miles north of Santa Fe, near Chama. She is struggling to help a local Hispanic community preserve a long heritage by keeping the weaving craft alive. We can be in the world only by being in a culture. The local Hispanic culture is under assault. While watching the movie *The Milagro Beanfield War,* one laughs and cries at the same time. There are heroes and villains in the story, of course, but in the real world of New Mexico and elsewhere much of the transformation seems the near inevitable consequence of cultures entwining and transforming one another. Fajitas, it seems, were invented in Texas, not Mexico. In New York, Chinese who fled Cuba have created Cuban-Chinese cuisine.

Will the emerging global civilization drift to homogeneity, as many suppose? Will we all speak English because the United States was powerful when television became widespread? Will we all love hamburgers? God knows, I do. But then I am a typical product of middle America.

Or will new cultural symbol strings sprout everywhere, created on every edge where two or more cultural traditions collide? Is the global civilization supracritical? If we find Cuban-Chinese cuisine, what else will we invent—say, on the frontiers of Islam and hard rock? Will we kill one another to preserve our ways of being in the world? What does tolerance demand of us when our ways of being in the world are swept into a whirlwind of change by the touch of someone else's memes beamed or e-mailed into our living rooms and studies?

For reasons I do not know, except perhaps that the image appeals to me, I find myself thinking about Per Bak and sandpiles again. I find myself suspecting that we will ever invent new cultural solar flares at the frontiers of bits of old colliding cultures. I find myself thinking about small and large avalanches of change propagating within and between the civilizations we have built in the past. I deeply fear the social havoc of dying ways of being in the world. People go to war for less. But one does find the idea of Cuban-Chinese cuisine and whatever might emerge from Islam and hard rock at least interesting. Maybe we need more of a sense of humor. Maybe we will know we are on our way when we can tell one another ethnic jokes because the mutual respect is so deep and the tolerance is so clear that laughter helps heal the remaining tensions.

Having listened to all these concerns, Scott Momaday returned to his own core thesis: we must reinvent the sacred in the modern world. Momaday's vision elicited from the four of us a strange sense: if a global civilization is emerging, we may be entering its heroic age, its age of creation. When Greek civilization collected itself on the Aegean shores, its early citizens constructed their own sustaining myths. We four "thought leaders," picked by some rather random process to assemble near Nambe New Mexico, found ourselves thinking that this emerging global civilization would have to invent its own new sustaining myths.

"Thought Leaders Gather on New Mexico Mountaintop and Advise World to Talk to Itself," quipped Walter Shapiro, as we ended our little meeting of the Gihon Foundation and went outside for lunch overlooking Michael Nesmith's gardens and the hills behind Nambe.

Reinventing the Sacred

Some 10,000 years ago, the last Ice Age began to falter. The ice sheets slowly retreated to the poles. In what later became the south of France, the Magdalenian culture—which had created the art in the caves of Font-de-Gaume and Lascaux, as well as upper Paleolithic flint blades, spears, and fishhooks of exquisite precision—faded. The large herds drifted northward. These ancestors drifted away, leaving the paintings that stun us today.

The bison and deer arched on these cave walls capture the sense of humanity's harmony with, reverence for, and awe of nature. No painting shows violence beyond images of hunting. One painting depicts two deer nuzzling. For some 14,000 years, these two have cared for each other on a stony curved wall in the Périgord.

Awe and respect have become powerfully unfashionable in our confused postmodern society. Scott Momaday said that we must reinvent the sacred. Our little meeting ended over a year ago. I lack Momaday massive frame, deeply resonant voice, and uncanny authority. Who am I to speak of these things? Another small voice. But has not our Baconian tradition, which celebrates science as the power to predict and control, also brought us a secular loss of awe and respect? If nature were truly ours to command and control, then we might well afford the luxury of contempt. Power corrupts, after all.

Friend, you cannot even predict the motions of three coupled pendula. You have hardly a prayer with three mutually gravitating objects. We let loose pesticides on our crops; the insects become ill and are eaten by birds that sicken and die, allowing the insects to proliferate in increased abundance. The crops are destroyed. So much for control. Bacon, you were brilliant, but the world is more complex than your philosophy.

We have presumed to command, based on our best knowledge and even our best intentions. We have presumed to commandeer, based on the availability of resources, renewable or not, that lay readily at hand. We do not know what we are doing. If Victorian England, standing astride an empire on which the sun never set, could in full good conscience see itself as the world's leader in persistent progress, where science meant the ensured betterment of mankind, can we see ourselves in such a way today?

We suspect ourselves. This is not new. Faust made his bargain. Frankenstein assembled his sad monster. Prometheus let loose fire. We have seen the fires we have lit spread beyond their intended hearthstones. We begin to know that proud humankind is still another beast, still embedded in nature, still spoken for by a larger voice.

If we find renewed concern about the untellable consequences of our own best actions, that is wise. It is not as though we could find a stance with either moral or secular certainty. We make our worlds together. All we can do is be locally wise, even though our own best efforts will ultimately create the conditions that lead to our transformations to utterly unforeseeable ways of being. We can only strut and fret our hour, yet this is our own and only role in the play. We ought, then, play it proudly but humbly.

Why try if our best efforts ultimately transform to the unforeseeable? Because that is the way the world is, and we are part of that world. That is the way life is, and we are part of life. We latter-day players are heritors of almost 4 billion years of biological unfolding. If profound participation in such a process is not worthy of awe and respect, if it is not sacred, then what might be?

If science lost us our Western paradise, our place at the center of the world, children of God, with the sun cycling overhead and the birds of the air, beasts of the field, and fish of the waters placed there for our bounty, if we have been left adrift near the edge of just another humdrum galaxy, perhaps it is time to take heartened stock of our situation.

If the theories of emergence we have discussed here have merit, perhaps we are at home in the universe in ways we have not known since we knew too little to know to doubt. I do not know if the stories of emergence we have discussed in this book will prove to be correct. But these stories are not evidently foolish. They are bits and pieces of a new arena of science, a science that will grow in the coming decades toward some new view of emergence and order in this far-from-equilibrium universe that is our home. I do not know if life began, as I have attempted to suggest, as an expected emergent collective property of the kinds of organic molecules that almost inevitably were formed on the early earth. Yet even the possibility of such collective emergence is heartening. I would rather life be expected in this unfolding since the Big Bang than that life be incredibly improbable in the timespan available. I do not know if the spontaneous order in mathematical models of genomic regulatory systems really is one of the ultimate sources of order in ontogeny. Yet I am heartened by a view of evolution as a marriage of spontaneous order and natural selection. I am heartened by the possibility that organisms are not contraptions piled on contraptions all the way down, but expressions of a deeper order inherent in all life. I am not certain that democracy evolved to achieve reasonable compromises between people with legitimately conflicting interests, but I am heartened by the possibility that our social institutions evolve as expressions of deep natural principles. "The Lord is subtle, but not malicious," said Einstein. We have only begun to invent the science that will account for the evolving emergent order I see out my window, from spider weaving her web, to coyote crafty on the ridgetop, to my friends and me at the Santa Fe Institute and elsewhere proudly hoping that we are unlocking some kinds of secrets, to all of you making your ways by your own best efforts and own best lights.

We are all part of this process, created by it, creating it. In the beginning was the Word—the Law. The rest follows, and we participate. Some months ago, I climbed to the first mountaintop I have been able to reach since my wife and I were badly injured in a car accident. I climbed to Lake Peak with Phil Anderson, Nobel laureate in physics and good friend at the institute. Phil is a dowser. I once was astonished to see him pull a forked twig off a tree and march across a hilltop holding it. I pulled off a forked twig and followed him. Sure enough, whenever his twig dipped toward the ground, so too did mine. But then, I could see him ahead of me. "Does it work?" I asked him. "Oh, sure. Half of all people can dowse." "Ever dig where your stick pointed?" "Oh, no. Well, once." We reached the peak. The Rio Grande Valley spread below us to the west; the Pecos Wilderness stretched out to the east; the Truchas Peaks erupted to the north.

"Phil," I said, "if one cannot find spirituality, awe, and reverence in the unfolding, one is nuts." "I don't think so," responded my dowsing, but now skeptical, friend. He glanced at the sky and offered a prayer: "To the great nonlinear map in the sky."

BIBLIOGRAPHY

Anderson, Philip W., Kenneth J. Arrow, and David Pines, eds. *The Economy as an Evolving Complex System*. Santa Fe Institute Studies in the Sciences of Complexity, vol. 5. Redwood City, Calif.: Addison-Wesley, 1988.

Axelrod, Robert. *The Evolution of Cooperation*. New York: Basic Books, 1984.

Bak, Per, and Kan Chen. "Self-Organized Criticality." *Scientific American*, January 1991, 46–54.

Dawkins, Richard. *The Blind Watchmaker: Why the Evidence of Evolution Reveals a Universe Without Design*. New York: Norton, 1987.

Eigen, Manfred, and Ruthild Winkler Oswatitsch. *Steps Towards Life: A Perspective on Evolution*. New York: Oxford University Press, 1992.

Glass, Leon, and Michael Mackey. *From Clocks to Chaos: The Rhythms of Life*. Princeton, N.J.: Princeton University Press, 1988.

Gleick, James. *Chaos: Making a New Science*. New York: Viking, 1987.

Gould, Stephen Jay. *Wonderful Life: The Burgess Shale and the Nature of History*. New York: Norton, 1989.

Judson, Horace F. *The Eighth Day of Creation: The Makers of the Revolution in Biology*. New York: Simon and Schuster, 1979.

Kauffman, Stuart A. *The Origins of Order: Self-Organization and Selection in Evolution*. New York: Oxford University Press, 1993.

Langton, Christopher G. *Artificial Life*. Santa Fe Institute Studies in the Sciences of Complexity, vol 6. Redwood City, Calif.: Addison-Wesley, 1989.

Levy, Steven. *Artificial Life: How Computers Are Transforming Our Understanding of Evolution and the Future of Life*. New York: Pantheon Books, 1992.

Lewin, Roger. *Complexity: Life on the Edge of Chaos*. New York: Macmillan, 1992.

Monod, Jacques. *Chance and Necessity*. New York: Knopf, 1971.

Nicolis, Gregoire, and Ilya Prigogine. *Exploring Complexity*. New York: Freeman, 1989.

Pagels, Heinz R. *The Dreams of Reason: The Computer and the Rise of the Sciences of Complexity.* New York: Simon and Schuster, 1988.

Pimm, Stuart L. *The Balance of Nature? Ecological Issues in the Conservation of Species and Communities.* Chicago: University of Chicago Press, 1991.

Raup, David M. *Extinction: Bad Genes or Bad Luck?* New York: Norton, 1991.

Shapiro, Robert. *Origins: A Skeptic's Guide to the Creation of Life on Earth.* New York: Summit, 1986.

Stein, Daniel, ed. *Lectures in the Sciences of Complexity.* Santa Fe Institute Studies in the Sciences of Complexity, Lectures vol. 1. Redwood City, Calif.: Addison-Wesley, 1989.

Wald, George. "The Original Life." *Scientific American*, August 1954.

Waldrop, M. Mitchell. *Complexity: The Emerging Science at the Edge of Order and Chaos.* New York: Simon and Schuster, 1992.

Winfree, Arthur T. *When Time Breaks Down: The Three-Dimensional Dynamics of Electrochemical Waves and Cardiac Arrhythmias.* Princeton, N.J.: Princeton University Press, 1987.

Wolpert, Lewis. *The Triumph of the Embryo.* New York: Oxford University Press, 1991.

INDEX

Abzymes, 120
Acetabularium, 158, 160
Acetylcholine, 137
Actin, 95, 97
Adaptation
  correlation length and, 196–98
  correlated fitness landscapes and, 169
  as hill climbing, 154
  long-jump, 192–99
  in patch procedure, 256
  of population, 221
Adaptive walks
  on fitness landscapes, 166–68, 184, 193
  between genotypes, 249–50
  population generations and, 226
  restricted, 205
Adenine, 35, 38
Adenosine triphosphate (ATP), 68
Adrenalin, 137
Affinity column, 141
Alanine, 134
Alchemy, 276–79, 286
Algorithmic chemistry, 277–79, 286
Algorithms
  complexity of, 153–54
  compressed, 248
  effectively computable, 21–22
  genetic, 246

Alleles, 162, 170, 222
Allolactose, 97, 99, 100, 104
Amgen, 133
Amino acids
  enzyme construction and, 44
  key–lock–key approach and, 139
  pairs of, 67
  peptide bonds in, 66
  polymers of, 37
  primitive earth atmosphere and, 36
  in proteins, 134, 142
Anderson, Phil, 169, 204, 235
AND function, 77, 80, 81, 103
AND gate, 77
Animal magnetism, 33
Annealing
  physical, 251–52
  simulated. *See* Simulated annealing
Antibody, 138, 145
  binding of, to antigen, 123, 142
  as candidate enzyme, 120
  catalytic, 120, 121, 147
  diversity of, 120–21, 127, 143, 146–47
  enzyme, 146
  generation of, 136
Antigens, 143
Applied molecular evolution, 132–41,
  147

Aqueous environment, 67
Aquinas, Thomas, 270
Arginine, 134
Aristotle, 5, 6
Arms race, 216
Arrow, Kenneth, 204
Arthur, Brian, 204, 279–80
Artifacts, 201, 246
  complex, 265
  evolution of, 242
  formation of, 191
  human crafting of, 202
Artificial life
  evolution of, 278
  extinction and, 238–39
Atmosphere, of primitive earth, 34, 35
Atoms of heredity, 162
ATP. See Adenosine triphosphate
Attenuation, of virus, 137
Attractors, 80, 107, 187
  defined, 78
  homeostasis, 83, 110
  order, 114
  order for free and, 79
  state cycle, 82, 100, 102, 107, 109,
    110
Auerswald, Phil, 204
Autocatalytic sets, 47, 72, 90, 147
  and catalysts, 63–64, 68
  catalytic closure and, 69, 72, 74, 90
  of catalytic polymers, 67
  coevolution of, 275
  compartmentalization of, 66
  competitors, mutualists and, 275
  computer simulation of, 64–65
  creation of, 49–50, 79
  defined, 49–50
  emergence of, 58, 61, 71
  endurance, likelihood of, 78
  evolution of, 73, 80, 278–79
  as functional integration models, 274
  hypersensitivity of, 74
  as level-0 organizations, 279
  of Lisp expressions, 278, 279
  metabolisms of, 288
  mutations in, 80
  orderly, 78
  origin of life and, 72
  as protocells, 86

reactions of, energizing, 66–69
reproduction of, 274–75
self-sustaining, 63, 64, 117
spontaneous, 74, 75
subsets of, 60–61
Autopoesis, 274
Autopoetic system, collective molecular,
  275
Avalanches
  of extinction, 212–13, 233, 236, 238,
    242–43
    measurement of, 233–34
  of novelty, 127–30
  of technological change, 241
Avery, Oswald, 38
Axelrod, Robert, 270–71

Bacteria, 32, 52, 71, 127, 138, 158
Bacterial ecosystems, species diversity
  and, 128
Bagley, Richard, 68, 72
Bak, Per, 28, 129, 235, 236, 243, 301
Banks, Peter, 266
Basin of attraction, 78, 83, 102, 110
B-cells, 95
Belosov–Zhabotinski reaction, 53, 54,
  66
Bergson, Henri, 33
Beta-galactosidase, 96
Beta-galactosidase gene, 96, 99
Biases, in control rules, 81
Big Bang, 19
Bilipid layer, 31–32
Bilipid membrane, 67, 150, 188
Bilipid membrane vesicles, 66, 72, 185
Binding, 123, 142, 144
Biological explosions, 116–18
Biological organization, law of, 108
Biosphere
  as collectively autocatalytic, 115
  ever-transforming, 281
  subcritical behavior of, 115
  supracritical behavior of, 123, 124
Biotechnology, 132, 133, 296
Biotechnology companies, 140
Blending inheritance, 162
Blind watchmaker, 201–2, 246
Blood, clotting of, 140
Boltzmann, Ludwig, 9

Bonding, 58
Boole, George, 77
Boolean algebra, 77
Boolean functions
  AND, 77, 80, 81, 103
  canalyzing, 103–6
  EXCLUSIVE OR, 104, 105
  of four inputs, 84–85, 89
  IF AND ONLY IF, 106
  noncanalyzing EXCLUSIVE OR, 106
  NOT IF, 100, 104
  OR, 77, 80, 81, 103, 104, 106
  rules of, 99, 103
Boolean hypercube, 165–66, 249, 254
Boolean networks
  attractors in, 79
  behavior field of, 102
  in chaotic regime, 82, 85, 88, 89, 91,
    103, 187, 223
  edge of chaos and, 213–14
  evolution of, 91
  family of, 81–82
  frozen components in, 219, 223
  homeostasis of, 110
  models of, 92, 99, 100, 223, 273
  NK, 228
  ordered, 85, 87, 187, 188
  perturbations in, 80
  state cycles in, 78
  wiring diagram of, 75–77
Bootstrapping, 288
Boundary, subcritical–supracritical, 129
Bounded rationality, 280
Brain, 137
Branching
  probability, of chain reactions, 214
  processes of, 125
  rate of, 202
Brenner, Sidney, 135
Buss, Leo, 158
Butterfly effect, 17

Cambrian explosion, 199–201
  coevolution and, 216
  diversity and, 15, 161, 192, 283
  multicelled organisms and, 93, 158,
    191
  origin of life and, 13, 14
  patterns of, 281

  technological evolution and, 202, 205
  universal law and, 27, 195
Canalyzing Boolean functions, 103–6
Candidate drugs, 136, 141
Candidate enzymes, 120, 124, 145
Candidate laws, 115, 235
Carbohydrates, 138, 143
Carbon dioxide, 34, 35, 207, 215
Cardiac arrhythmias, 54
Carnot, Sadi, 9
Catalysis, 69, 72, 144. *See also* Catalysts
  defined, 49
  open thermodynamic systems and, 50
  reaction networks and, 58, 61, 120, 129
Catalysts
  antibodies as, 120, 121, 147
  in autocatalytic sets, 68
  biological explosions and, 116, 117
  chemical reaction rate and, 49, 51
  conditions for, 58
  defined, 49
  dual role of, 59
  for randomly chosen reactions, 123
  in reaction networks, 60
  reaction process of. *See* Catalysis
  rules of, 63
Catalytic closure
  autocatalytic sets and, 69
  defined, 50
  molecular systems and, 275
Catalytic task space, 144
Catalyzed reaction subgraph, 60, 64
Catastrophic variation, 154
Cech, Thomas, 40, 41
Cells, 10–11, 52, 89, 158
  archaic forms of, 31
  daughter, 34, 72, 73, 110
  differentiation of, 24, 93–96
  division cycle of, 37, 93, 107
  eukaryotic, 157, 158, 215, 223
  fossil, 10, 11, 12, 31, 159
  of free-living systems, 21
  of immune system, 95
  membranes of, 150
  metabolic networks in, 86
  metabolism of, 124
  muscle, 24
  nerve, 94, 95
  nonequilibrium systems and, 22

Cells (*continued*)
  primitive, 35
  receptors on, 146
  skeletal-muscle, 95
  somatic, 37
  subcritical behavior of, 126
  types of, 94
Central directing agency, 274
Central dogma of developmental
    biology, 95, 106
Chain reactions, 125, 129
Chaitin, Gregory, 155
Chance and Necessity (Monod), 97
Chaos, 79, 80
  Boolean networks and, 82, 85, 88, 89,
    187
  edge of, 86–92
  theory of, 17, 23, 210
Chaotic regime, 73–74, 247
  Boolean network in, 82, 85, 88, 89,
    91, 103, 187, 223
  network in, 87
  sensitivity of, to change, 88
Chemical creation myth, 54–58
Chemical equilibrium, 51
Chemical reaction, 24, 60, 119
  rates of, 67, 266
  systems, 122
Chemostat, 184, 278
Chloroplasts, 158, 215
Chromosomes, 37, 94, 162
  diploid, 180
  in meiosis, 180–81
  paternal, 95
  recombination of, 181–82
Chunking, 265
Church, Alonzo, 284
Clades, bottom-heavy, 200, 201
Classes, 199
Cleavage reactions, 62, 68
Cloning, 132
Coascervates, 35, 66
Coevolution, 240, 246, 264
  of autocatalytic sets, 275
  biological evolution and, 216–17
  in economic/cultural systems, 217
  edge-of-chaos and, 27–28
  evolution of, 221–24, 230
  fitness landscapes and, 208

game theory and, 217–21
  mutualism and, 215–16
  organismic, 282
  of patches, 256–57
  symbiosis and, 215–16
  technological, 279–84
  technological evolution as, 280
Collective autocatalysis, 274–75
Collectively autocatalytic sets. *See*
    Autocatalytic sets
Combinatorial explosions, 248, 249
Communication, receiver-based, 267–69
Community, 209–15
  fitness of, 214–15
Community landscape, 212
Community-assembly simulation
    studies, 214
Compartmental systems, 72, 73
Competitive advantage, 246
Complementarity, 282, 287, 290–91
Complexity, 87, 117
  boundaries and, 265
  coordination of, 90
  diversity and, 296–97
  laws of, 90, 114, 124
  sciences of, 246
Complexity catastrophe, 194–95
Complex systems
  coordination of, 90
  evolution of, 155, 157
  gradualism and, 151, 152
  natural selection and, 155
  properties of, 18–19
Computation, theory of, 21–22, 23
Computer programs, 152
  compressed, 155
  serial, 152–53
Computer simulations, 64, 68, 91, 117,
    176–77, 225
Conflicting constraints, 264
  increasing, 258–59
  in rugged landscapes, 258
Conflict-laden problems, 250–52, 269,
    270
Control rules, 81
Copernicus, Nicolaus, 5
Correlation length, of fitness landscapes,
    193
Cosmogenesis, 50

Coupled fitness landscapes, 224–35
Creation science, 6
Cretaceous era, 14, 238
Cribiform plate, 42–43
Crick, Francis, 36, 38
Critical molecular diversity, 64, 65
Critical value branching processes, 129
Cross-coupling, 173
Crypts of Leberkuhn, 107
Cullum, Lee, 298, 301
Cytosine, 35, 38

Darwin, Charles, 149, 150
    blind watchmaker vision of, 201–2
    gradualism and, 151–52, 154, 161
    natural selection and, 6, 7, 13, 22, 26,
        72, 73, 80, 85, 92, 97, 161, 162
Daughter cells, 34, 72, 73, 110
Daughter neutrons, 126
Daughter species, 199
Dawkins, Richard, 277, 300
Decentralization, 245–46, 266
Dehydration, 68
Democracy, 28, 299
Deoxyribonucleic acid. *See* DNA
Derrida, Bernard, 84, 88
Dickinson, Emily, 91, 247, 253
Differential equation, 209
Digestion, 123
Dipeptides, 67
Diploids, 180
Discount factor, 294
Dissipative structures, 21, 53
Diversity
    of bacterial species, 127
    of biosphere, 115
    Cambrian explosion and, 15, 161,
        192, 283
    of cell types, 110
    complexity and, 296–97
    critical molecular, 64, 65
    increased, 283
    molecular, 62, 113–14, 122, 123, 128
    of polymers, 117
    of species, 113, 129
    technological, 292–94
DNA, 24, 72, 86, 95, 129, 145
    applied molecular evolution and, 132,
        133, 135, 140, 142

catalysis of, 129
circles of, 216
double helix of, 36, 38
genetic information in, 24, 95
protein encoding by, 132
random sequences in, 136
regulatory sites, 145
structure of, 35, 36, 38, 186
DNA space, 134
Dreisch, Hans, 33, 34, 94
Drugs, discovery of, 135
Dynamical order, 74
Dynamical systems, 75, 187

Earthquakes, size distribution of, 236
Economic systems, 22, 27–28
    evolution of, 169
    global changes in, 245–46
    webs of, 281, 282, 295, 297
Economy
    diversity in, 294
    ecology and, 275
    ever-transforming, 281
    growth theories of, 294–95
    mushroom webs of, 288
    takeoff of, technological coevolution,
        grammar models and, 289–98
Ecosystems, 22, 73, 114
    age of mortality of, 242
    community behavior of, 209–10
    evolutionary stable strategy of, 226
    evolution of, 169
    as mushroom webs, 288
    progenitors of, 12
    self-tuned to the edge of chaos, 234–35
    subcritical behavior of, 129
    as web of coevolving agents, 240
Ectoderm, 94
Ectodermal cells, 110, 111
Edge of chaos, 26–29, 86–92, 246, 257–64
    in Boolean genomic networks, 213–14
    as complex regime, 103, 223
    evolutionary stable equilibria and,
        229–30
    patching systems and, 266
Ediacrian period, 161
Egg, 37, 95
    fertilization of, 71, 93
    maturation of, 180

Eigen, Manfred, 184, 276
Einstein, Albert, 66
Elan vital, 33, 48
Electric potential, 33
Ellington, Andy, 140, 141
Embryos, 33, 110
Emergence, theory of, 23–24
Endergonic reactions, 68
Endoderm, 94
Endosymbionts, mitochondrial, 223
Endosymbiotic alliance, 215
Energy, 66, 68
    barriers, 251
    landscape, 257
    metabolism, 158
    minimums, rugged landscapes and,
        25–62
Entelechies, 34
Entropy, 9
Enzymes, 44, 71, 73, 75, 97, 128, 145
    activation of, 106
    binding sites on, 287
    in catalysis, 49, 51, 144
    catalytic activities of, 74
    chemical reaction rate and, 67
    complementarities, 283
    conformation of, 106
    functions of, 36, 59
    induction of, 96
    molecular feedback and, 74
    proofreading by, 40, 41
    proteins and, 35
    substrates and, 277, 285–86
Epistatic coupling, 170–71, 224
Epitopes, 137, 138
EPO. See Erythropoietin
Equilibrium systems, 9–10, 20, 51
Ergodic hypothesis, 9–10
Error catastrophe, 40, 41, 152, 161, 184,
    188
Erythropoetin (EPO), 133
Escherichia coli, 86, 95, 96, 300
ESS. See Evolutionary stable strategies
    (ESS)
Estrogen, 146
Eukaryotic cells, 158, 215, 223
Evolution, 4, 7, 19, 23, 72, 90, 98, 161
    biological, 26–27
    of coevolution, 221–24

    conflicting design criteria and, 14
    fitness landscapes and, 166
    fluctuations and, 164
    laws of, 192
    by mutation, 180
    by natural selection, self-organization
        and, 185–89
    organismic, 282
    on rugged landscapes, 175–80
    by selection, 180
    successful, 186–87
    theory of, 6, 25
Evolutionary stable strategies (ESS),
    234, 258, 264
    coevolution and, 226
    defined, 220–21
    disruption of, 231–32
    equilibrium of, 241–42
    freezing of, 228–29, 253
    ordered regime, 223
    unfreezing of, 235
EXCLUSIVE OR function, 104, 105
Exergonic reactions, 68
Exons, 41
Extinction, 14–15, 27
    avalanches of. See Avalanches: of
        extinction
    in community assembly, 212–14
    events, controlling, 232–33
    of species, 211
Extraembryonic membranes, 93

Farmer, Doyne, 68, 72
Fibonacci series, 151, 185
Finite size effects, 238
Fischer, Emil, 33
Fisher, Ronald A., 162
Fitness
    changes in, correlation length and,
        196–98
    highest, 228, 231
    landscapes. See Landscapes
    natural selection and, 149, 158
    peaks, 26–27, 248–49, 253
    structure of, 162–63
    values, 234
Flexibility, 91
Fontana, Walter, 276, 277–78
Food webs, 211

Formal undecidability, 288
Founder set, 288
Founders of families, 199, 200
Free-living systems, 21
Functional integration, autocatalytic sets
    and, 274
Fungi, 11

Galilei, Galileo, 5, 16
Game theory, 217–19
Gases, 35
Genera, 199, 239–40
Generation, species in, 231
Genes, 71, 94, 95, 111
    cloning of, 132
    epistatic coupling of, 170–71, 224
    molecular structure of, 36
    nude, 39
    splicing of, 132
    structural, 96
Genetic algorithm, 246
Genetic circuits, 96, 101
Genetic drift, 129
Genetic information, 37, 41
Genetic regulatory circuits, 99, 103
Genome
    autocatalytic set evolution and, 73
    complexity of, limit on, 184
    of humans, 102, 115, 124
    random change in, 154
    regulatory networks. *See* Genomic
        regulatory networks
    spontaneous order and, 106
Genomic regulatory networks, 85, 188,
    237
    model of, 99, 100
    NK model of, 170–73
    ontogeny and, 24–25, 33, 102
    in ordered regime, 106–7
    phase transitions in, 257
    principles of, 103
Genotypes, 73, 96, 151
    fitness of, 165, 166–67, 172
    global optimum of, 173
    neighboring, 165–66
Genotype spaces, fitness landscapes
    and, 163–69
Germ layers, 94
Germ line, 274

Germ plasm, 95
Giant component of random graph, 56
Gibbs, Josiah Willard, 9
Gihon Foundation, 298, 302
Global civilization, 298–302
Global positioning systems (GPS), 297
Glycine, 134
God, 4, 5, 6, 29, 69, 180–83, 201
Gödel, Kurt, 288–89
Goodwin, Brian, 273, 275
Gould, Stephen Jay, 13
GPS. *See* Global positioning systems
Gradualism, 151, 154, 188
    correlated fitness landscapes and, 169
    ideal, 174
    selection limits and, 183
Grammar laws, 293
Grammar models, 287, 291–92, 294
Grammar space, 287
Grammar table, 291
Great Red Spot, on Jupiter, 20–21
Green, Paul, 151
Grypanice spiralis, 159
Guanine, 35, 38

Haldane, J. B. S., 162
Halifax, Joan, 297
Hard combinatorial optimization
    problems, 248
Hardy, George H., 162
Hemoglobin, 24, 36, 95
Heredity, 37
Heritable variation, 72, 92
Hexapeptide, 139–40
Hierarchies, of organizations, 279
High-energy bonds, 68
History, 299
Holism, 68
Homeostasis, 74–80, 79, 83, 86, 91, 92,
    110
Homeostatic molecular coordination, 126
Homeostatic stability, 79
Homo habilis, 93, 131, 148, 275–76, 280
Homo ludens, 131, 147, 148
Homo sapiens, 93, 131, 148, 191, 265
Homunculus, 34, 94
Hormone ligands, 146
Host–parasite systems, coevolution of,
    216

Hoyle, Sir Fred, 44, 47
Human genome, 102, 115, 124
Human genome project, 133
Humans
  intermediate metabolism in, 125
  problems facing, 4–5
Humpty Dumpty effect, 211
Hydrochloric acid, 95
Hydrogen, 32, 34, 37
Hyperparasites, 238

Idealizations
  of Boolean functions, 103
  usefulness of, 74–75
IF AND ONLY IF functions, 106
Immune system, 95, 136, 143
Incompleteness theorem, 288–89
Infant mortality
  in artificial life, 240–41
  high, 241
  in marine life, 240
Inner cell mass, 93
Innovation, 280
Inorganic atoms, 32
Insulin, 136
Introns, 41
Invisible hand, 19, 246, 247, 262
Isotropic fitness landscapes, 177

Jacob, François, 95–97, 98, 99
Jacobs, Jane, 295

Kant, Immanuel, 69, 270, 275, 279
Kepler, Johannes, 5–6, 16, 22
Key–lock–key procedure, 137, 138, 139
Koza, John, 157
K values, 232, 234, 235

Lactose, 95–96, 97, 101
Lambda calculus, 135, 277, 284
Landscapes, 161, 165, 223
  adaptive processes on, 204
  adaptive walks on, 166–68, 184, 193
  building of, 170, 171–73
  changes in, 208
  community on, 212
  conflicting constraints and, 173, 187
  correlated, 169–75, 188, 198–99
  correlation length of, 193

coupled, 224–35
families of, 170
God's-eye view of, 180–83
isotropic, 177
jumping across, 192–99
life on, 157–63
natural selection and, 182–83
NK model of. See NK landscape model
nonisotropic, 177
random, 166, 167, 169, 174, 178, 204,
  248
rugged, 187, 208, 231, 240, 250
  conflicting constraints and, 258
  energy minimums of, 25–62
  evolution of, 175–80
  general features of, 200
  technological evolution on, 201–3
  search for excellence and, 247
Langton, Chris, 90
Learning curves, 192, 203–6, 280,
  295–96
Leftist Italian limit, 258, 271
Level-0 organizations, 279, 300
Level-1 organizations, 279, 300
Life
  experimental creation of, 147–48
  networks of, 48–54, 75
  origin of, 4, 19, 27–28, 32, 43–44
    central idea of, 61–66
    collective autocatalysis and, 274–75
    critical diversity of molecules and,
      63–64
    crystallization, 43–45, 58, 64, 117
    as emergent phenomenon, 24
    laws of, 16–23
    precellular, 11
    problem of, 34
    theories of, 32, 34, 58, 69, 111–12
Life-forms, earliest, 10–12, 14
Ligation reactions, 68
Limit cycle, 210
Linnean taxonomy, 7, 13, 199
Lipids, 112, 185, 186
Lipid vesicles, 112
Lisp, 277, 278–79, 283
Lobo, José, 204
Logarithmic scales, 108
Logic, of patches, 247–52
Long-jump adaptation, 192–99

Lotka, A. J., 209–11
Lotka–Volterra equations, 210
Lower taxa, 191
Lysine, 134

McCaskill, John, 276
McCulloch, Warren, 273
Macready, Bill, 91, 196, 204, 247, 253, 254
Macroeconomics
  models of, 289
  theories of, 295
Mahler, Gunter, 3
Maleability, 73
Match catalyst rule, 63
Maturana, Humberto, 274
Maxwell, James Clerk, 33
Meiosis, 37, 180
Memes, 277
Mendel, Gregor, 37, 162
  laws of, 37, 94, 162
Mesoderm, 94
Messelson, Matthew, 38
Metabolic reaction graph, 58
Methane, 34, 35
Mill, James, 5, 271
Miller, Stanley, 35, 36
Misspecification, 266–67
Mitochondria, 158, 215, 223
Mitosis, 37
Molecular creativity, 132
Molecular feedback, 74
Molecular novelty, 132
Molecule recognition, 142
Momaday, N. Scott, 4, 298, 302
Monod, Jacques, 71, 95–97, 98, 99
Monomers, 39
Morgenstern, Oskar, 217
Morphogenesis, 93, 94
Mortality curves, 241
Mosaic development, 33–34
Mushroom webs, 288
Mutations, 40, 80, 91
  in autocatalytic sets, 80
  in early development, 200–201
  on fitness landscape, 166
  genetic algorithm and, 246
  to genome, 151, 222
  gradualism and, 169, 183
  large number of, 193
  in late development, 200, 201
  in network, 83
  random, 129, 133
  rate of, 184
  redundancy and, 156
  revertant, 137
  source of, 161–62
  viral, 216
Mutualism, coevolving, 215
Myosin, 24, 95, 97

Nash, John, 218
Nash equilibrium, 218, 220, 223, 264
Natural selection, 112, 154, 163, 201
  complex systems and, 155
  culling fitter variants and, 164–65
  Darwin and, 6, 7, 13, 22, 26, 72, 73, 80, 85, 92, 97, 161, 162
  efficacy of, 150, 165
  evolution and, 304
  fitness and, 149, 158
  fitness landscapes and, 182–83
  genetic algorithm and, 246
  at individual organism level, 208–9
  limits to, 152, 183–85, 188, 194
  self-organization and, 166, 185–89
  as sole source of order, 97–99
Nelson, Dick, 280
Nerve cells, 94, 95
Nervous system, 137
Nesmith, Michael, 302
Networks
  of life, 48–54, 75
  parallel-processing, 186
Neumann, Kai, 230, 234, 239, 241
Neurotransmitters, 137
Newton, Sir Issac, 6
  laws of motion of, 18
Niches, 231, 240, 282
  autocatalytic creation of, 283
  creation of, 281, 287
  defined, 280
NK landscape model, 170–73, 183, 187, 204, 222
  conflict-laden problems and, 247–48
  correlation length of, 193
  epistatic interactions and, 224–25
  increasing range of, 259, 262

NK landscape model (*continued*)
  lattice and patch form of, 267
  long-jump adaptation in, 192–99
  patch procedure in, 255–67
  phase transition in, 262–64
  receiver-based communication in,
    268–69
  square lattice form of, 253–54
Noah's vessel experiment, 122–27, 128,
  130
Noncanalyzing EXCLUSIVE OR function,
  106
Nonconvergence, 265
Nonequilibrium systems, 20, 82, 187
  chemical, 51, 54
  free-living, 21
  open, 52
  ordered, 20–21. *See also* Dissipative
    structures
Nonisotropic fitness landscapes, 177
NOT IF function, 100, 104
NOT IF rule, 103
Novel molecules, 124, 126, 128, 131
Nuclear chain reaction, 114
Nuclei, 37
Nucleotide, 35, 36, 38
  bases of, 39, 41

Oligonucleotides, 119–20
Ontogeny, 92, 93, 95, 99, 102, 201
  developmental program of, 274
  genomic regulatory networks and,
    24–25, 33, 102
  order of, 103
  properties of, 107
Oparin, Alexander, 35, 66
Operator, 96, 99
Operator DNA sequences, 97
Operator site, 97, 104
Optimization problems, hard
    combinatorial, 248
Order, 71, 80, 83, 187
  biological, 24
  of ecosystems, 149
  equilibrium systems and, 9–10
  forms of, 20–21
  low-energy equilibrium forms of, 20,
    22
  of natural world, 8

nonequilibrium, dissipative structures
    of, 20–21, 22
  of ontogeny, 103
  in organisms, 150
  requirements for, 80–86
  selection and, 97–99
  source of, 150, 152
  spontaneous. *See* Spontaneous order
  sustained nonequilibrium structure
    of, 20–21
Order–chaos axis, 91, 228, 229–30
Ordered regime, 90, 91, 111, 247
  frozen components in, 223
  genomic networks and, 188
  Stalinist limit in, 255–57, 258
Order for free, 71, 75, 79, 89, 91, 92,
  99
Ordovician era, 13, 199
OR function, 77, 80, 81, 103, 104, 106
Organelles, 158
Organic molecules, 119
  indexes of, 115
Organisms, 71, 246
  coevolution of, 282
  collectively autocatalytic metabolisms
    of, 288
  evolution of, 242, 282
  fitness of, 170
  foraging by, 179
  hierarchical categorization among,
    199
  multicelled, 158, 161
  Cambrian explosion and, 93, 158,
    191
  sexual, 180
  single-celled, 158
  speciation of, 240
  as wholes, 69
Organizations, 246
  hierarchies of, 279
  level-0, 279, 300
  level-1, 279, 300
  patches. *See* Patches
Organs, 94
Orgel, Leslie, 41, 48
Origins (Shapiro), 36, 44
Origins of Order (Kauffman), 84, 194
Oster, George, 142, 143, 144
Oxygen, 32, 215

Packard, Norman, 68
Paleozoic era, 126
Paley, William, 150, 201
Palmer, Richard, 169
Paradise, loss of, 3–4, 5
Parallel-processing networks, 186
Parasites, 238
Pasteur, Louis, 32
Patches
   boundaries of, 268
   logic of, 247–52
   nonconvergence and, 265
   optimal, 265–66, 267
   optimum size distribution of, 262–64
   and politics, 270–71
   possibilities of, 264–67
   procedure, 252–57
   receiver-based communication and,
      268–69, 270, 271
   size of, energy minimums and, 258
Patching systems, edge of chaos and,
   266
Pauli, Wolfgang, 98
Peptide bonds, 66, 68
Peptides, 66, 140, 141, 143, 145
   erythropoietin, 133
   key–lock–key approach and, 136
   libraries of, 135
   novel, 138
   random, 136
Perelson, Alan, 142, 143, 144
Periodic motion, 18
Permian extinction, 13, 199–201, 238
Perturbations, 89, 91, 164
Phanerozoic, 237
Pharmaceutical companies, 140, 145
Phase transitions, 56, 57, 64, 80, 187,
   223
   biological explosions and, 116–18
   to collective autocatalysis, 274
   life emerges as, 62
   between order and chaos. *See* Edge of
      chaos
   patch procedure and, 253
Phenotype, 73
Photosynthesis, 158
Phyllotaxis, 151, 185
Phylum, 199, 200
Physical annealing, 251–52

Pimm, Stuart, 211*f*12, 214, 215
Placenta, 93
Planning horizon, 291–92, 294
Plasmids, 216
Plastein reaction, 68
Pleuromona, 42, 69
Poised systems, 236–37
PolyG, 40
PolyG decamers, 39
Polymerase, 41
Polymers, 62, 63, 66, 132, 142
   diversity of, 128
   formation of, 67–68
   large, 115, 129
   model, 117
   ratio of possible reactions to, 65
   to reproduce, autocatalytic systems
      of, 68
Population
   generations, 225–26
   trajectories, 212
Post, Emil, 284
Post-Darwinian biologists, 150
Postextinction rebounds, 200
Post grammar, 284
Power-law distributions, 29, 203–5, 243
   avalanches of extinction and, 212–14,
      233–36
   critical value branching processes
      and, 129
   of extinction events, 237
   self-organized criticality and, 28–29,
      235–43
Prebiotic chemical systems, 24, 27
Precambrian, 158
Predator–prey relationships, 210, 211,
   216
Preformationists, 34, 94
Prigogine, Ilya, 53
Prisoner's Dilemma, 217–19, 220, 223,
   270
Probability
   of catalysis, 124
   models of, 147
Production substitutes, 282
Products, 75
Progeny, 158
   expected number of, 125
Proteins, 35, 71, 73, 95, 140, 142

Proteins (*continued*)
  binding of, 96
  as candidate enzymes, 120
  in catalytic reactions, 129
  digestion of, 68
  dual role of, 59
  as epitopes, 138
  genetic coding for, 72
  genetic encoding of, 115, 124
  as genetic material, 36
  libraries of, 145
  number of, 133–34
  random, 124–25, 143, 147
  as repressor, 97
  structure of, 35
  synthesis of, 95
Protocells, 35, 73, 86, 89, 114
Protoearth, 31, 113
Prozac, 137
Pyrophosphate, 68

Quantum theory, 23

Random catalyst rule, 63
Random chemistry, 145–47
Random grammars, 284–87
Random graph, 54–58
Random motion, 251
Random systems, 18
Random variations, 7, 8, 98, 201
Rational morphologists, 6–7
Raup, David, 208, 237, 239
Ray, Tom, 237, 238, 239
Reaction energy, 66–69
Reaction graph, 58, 59, 62
Reaction networks, 58–61
Receiver-based optimization,
    267–71
Receptors, 136, 137
Recombination
  chromosomal, 181–82
  genetic algorithm and, 246
Red blood cells, 24, 94, 95
Red Queen effect, 216, 221–22, 224,
    227, 231, 253, 257–58
Reductionism, 16, 33
Redundancy, 153, 154, 188
Redundant program, 156
Reed, John, 204

Repressor, 96, 97, 100
Ribonucleic acid. *See* RNA
Ribosomes, 41
Ribozyme polymerase hypothesis, 41
Ribozyme RNA polymerase, 278
Ribozymes, 40, 41, 42, 51, 59, 63, 141
Richter scale, 236
RNA, 102, 116, 145, 186
  applied molecular evolution and, 133,
      140–42
  catalysis by, 129, 140
  dominant nude, 69
  dual role of, 59
  folding of, 276
  genetic information on, 72
  in genomic regulatory network,
      102–3
  landscapes, 277
  libraries of, 135
  messenger, 39, 96
  nude replicating, 42
  ontogeny and, 24–25
  random sequences, 136, 140, 143
  replication of, 47, 80
  sequences, 63, 141
  structure of, 35, 36, 39, 48
  transcription of, 95
  transfer, 40
RNA space, 134
Robustness, 187–89
Robust systems, 157
Runnegar, Bruce, 158, 161

Scaling relation, genes to cell type,
    109–10
Schenkman, José, 295
Schumpeter, Joseph, 240, 243, 279
Schuster, Peter, 184, 276
Science
  of complexity, 4–5, 8
  of creation, 6
  theoretical, 17–18
Selection. *See* Natural selection
Selection-is-the-only-source-of-order
    view, 150
Self-organization, 26, 71, 99, 112, 246
  Belosov–Zhabotinski reaction and,
      53–54
  emergence of life and, 43

history of life and, 150
natural selection and, 166
as order for free, 71, 75, 79, 89, 91, 92, 99
phyllotaxis and, 151, 163, 246
power of, 45, 92
robustness and, 189
selection and, 98, 163, 185–89
spontaneous, 152
Self-organized criticality, 28–29, 235–43
Self-repeating pattern, 151
Self-reproduction, 72, 73, 92, 114
by earliest molecular systems, 31
types of, 278–79
Sepkowski, Jack, 239
Serotonin, 137
Sex, 180, 181
Shape space, 142
Shapiro, Robert, 36, 44
Shapiro, Walter, 298, 301
Shell, Karl, 204
Sigmoidal curve, 57
Signaling molecules, 136
Simulated annealing, 248–49, 250, 252, 269
Single-mutant neighbor, 231
Smith, Adam, 19, 208
Smith, John Maynard, 219–20, 299
Social-planner model, 292–94
Speciation, 14–15
Species, 208
in biosphere, 115
diversity of, 129
epistatic connections among, 228–29
extinction of, 211
interactions of, 230–31
molecular, explosion of, 116
patches as, 253
reservoir, 211
younger, chance of extinction and, 242
Sperm, 37, 93, 95, 180
Spin-glasses, 169–70
Spirituality, 302–4
Spontaneous generation, 34, 45
Spontaneous order, 8, 18, 61, 91, 188
arises, 75
genome and, 106
natural selection and, 71, 304

of ontogeny, 99–111
sources of, 149, 186
Spontaneous reactions, 58, 60, 63, 66, 72
Spores, 21
Square-root relation, of cell types to DNA, 108–9
Stability, 73, 91
Stahl, Franklin, 38
Stalinist limit, 255–57, 258, 260–61, 262, 271
State cycle, 77–79, 82
State-cycle attractors, 82, 100, 102, 107, 109, 110
States, sequence of, 77
State space, 75, 106
Statistical mechanics, 9, 18
Steady state
isolated, 78
single, 78
Stein, Dan, 169
Stromatoliths, 12
Structural stability, 188. *See also* Redundancy
Subcritical behavior, 117, 145, 293, 294
Subcritical–supracritical boundary, 126, 129, 130, 132, 148
Substitution, laws of, 285, 286
Substrates, 58, 60, 74, 75
Subtle medicines, 133
Sudden death, 54
Sugars, 207
Suppressor oncogenes, 140
Supracritical behavior, 114–16, 122
of biosphere, 145
boundary of, 129
economic, 294
Supracritical explosion, 127, 128, 145, 157
Supracritical molecular diversity, 123
Supracritical reaction systems, 118–22
Surface reactions, 67, 68
Sutherland, Stuart, 98
Symbiosis
coevolution and, 215–16
mutual consistency and, 223
Symbol strings, 277, 278, 283, 298
collectively autocatalytic set of, 287
cultural, 301

Symbol strings (*continued*)
 founder set of, 288
 Post grammar of, 284–87
 technological coevolution and,
   289–91
Symmetry, 151
Synthesis, free energy to drive, 68
Szostak, Jack, 140, 141, 278

Tang, Chao, 28, 129, 235, 236
T-cells, 216
Technology
 coevolution of, 279–84
  economic takeoff and, 289–98
 evolution of, 4, 14, 15, 27–28, 191,
   205
  laws of, 192
   learning curves and, 192, 203–6
   on rugged landscapes, 201–3
  trajectories of, 203, 204–5
Temperature, 252
 chemical reaction rates and, 266
 as measure of thermal vibration,
   251
Template, replication of, 47, 72
Temporal Organization of Cells, The
   (Goodwin), 273
Tetrapeptides, 67
Thermal vibration, 251
Thermodynamics, 66, 92
 second law of, 9–10, 71
 systems
  closed, 50, 51
  open, 82, 83, 92, 208
Thomas, Dylan, 207
Thrombin, 140
Thymine, 35, 38
Tierra, 238–39
Tissues, 94
Tit for tat strategy, 219
Total quality management, 246
Toxins, 128, 134
Toy problem, 54–58
Traits, 170, 224–25
Trajectory, 77, 100
Transcription, 96, 97
Transitional region, 90
Transition state, 51, 52, 120, 123, 144

Traveling Salesman problem, 249, 250,
   253
Tripeptide, 67
Trypsin, 59, 68
Turing, Alan, 22, 284
Turing machines, 276–77, 284, 285

Uncatalyzed reaction, random, 72
Universal computational systems, 22
Universality classes, 283–84, 287
Universal molecular toolboxes, 141–45
Universe, nonequilibrium, 19–20
Uracil, 35
Uranium, nucleus of, 114
Urea, synthesis of, 33
Urey, Harold, 35

Vaccines, 137, 145
 creation of, 138
 novel, 138
 production of, 137
Van Valen, Lee, 216
Varela, Francisco, 274
Variants, fitter, 204–5
Variations, catastrophic, 154
Vendian period, 158, 161
Verela, Maria, 301
Viruses, 21, 42, 216
 attenuation of, 137
 hepatitis, 137
 HIV, 216
 polio, 137
 self-assemblying, 185, 186
Volterra, V. J., 209–11
Von Neumann, John, 217

Wald, George, 44
Warnock, Geoffrey, 98
Water, 31, 37, 66, 68, 207
 on primitive earth, 35
Watson, James, 36, 38
Watson–Crick base pairing, 39, 47, 48,
   50, 63, 135
Webster, Gerry, 274
Weinberg, Steven, 16
Weinberg, W., 162
Weininger, David, 115

Weisbuch, Gerard, 84, 88
Weismann, August, 274
When Time Breaks Down (Winfree), 53
Whirlpool dissipative system, 187
White cells, 94
Wickramasinghe, N. C., 44, 47
Wiesenfeld, Kurt, 28, 129, 235, 236
Winfree, Arthur, 53, 54
Winter, Sidney, 245, 280
Wolpert, Lewis, 158

Wonderful Life (Gould), 13
Wood, Larry, 267–68
Wright, Sewall, 162

Zygote
  deletions and, 34
  differentiation of, 24, 33, 93–94, 110
  genetic regulatory network in, 37, 95,
    98, 103
  genomic networks in, 26